CONQUERING THE ELECTRON

CONQUERING THE ELECTRON

THE GENIUSES, VISIONARIES, EGOMANIACS, AND SCOUNDRELS WHO BUILT OUR ELECTRONIC AGE

Derek Cheung and Eric Brach

ROWMAN & LITTLEFIELD
Lanham • Boulder • New York • London

Published by Rowman & Littlefield
A wholly owned subsidiary of The Rowman & Littlefield Publishing Group, Inc.
4501 Forbes Boulevard, Suite 200, Lanham, Maryland 20706
www.rowman.com

16 Carlisle Street, London W1D 3 BT, United Kingdom

Originally published in Chinese by Commonwealth Publication, Taiwan.

British Library Cataloguing in Publication Information Available

Library of Congress Cataloging-in-Publication Data Available
ISBN 978-1-4422-3153-5 (cloth : alk.paper)
ISBN 978-1-4422-3154-2 (electronic)

∞™ The paper used in this publication meets the minimum requirements of American National Standard for Information Sciences—Permanence of Paper for Printed Library Materials, ANSI/NISO Z39.48-1992.

Printed in the United States of America

This book is dedicated to my father,
Mr. Y. C. Chang,
and to my teacher,
Professor Gerald L. Pearson.
—Derek Cheung

CONTENTS

INTRODUCTION

Look at your cell phone. If you're like most people, your phone is not really just a phone. In its tiny, light, and elegant package, it provides multiple incredibly powerful features and capabilities. Far beyond just making calls, today's phone can simultaneously function as a web-surfing terminal, an eBook reader, an electronic game center, a GPS navigator, a high performance camera, and a music player with its own personalized library—to say nothing of the millions of other applications it can execute and perform. It runs via innovative hardware and software that are the results of many years of accumulated scientific and technological development, and whose existence is made possible by human beings' ability to totally master the flow of electrons.

We may not realize it but, with the swipe of our fingertips across our phones' screens, we are literally commanding billions of tiny electrons across silicon chips inside the phone, which are designed like microscopic maps of giant cities, interconnected by the bridges and tunnels of laminated wiring. Following our specific software instructions, these electrons tirelessly do massive amounts of work for us, from performing complex mathematical and logic functions to storing and retrieving information. At the same time, the compatriot of the electron, the radio wave, shuttles around in space, wirelessly delivering our coded messages and other data at the speed of light. These almost magical capabilities can be traced to the power of the tiny electron. ***This prompts inevitable questions: How did mankind conquer the electron? What was this discovery journey like? Who were the main contributors who led the way? And what have they accomplished?***

This book presents the history of the conquest of the electron and how mankind has harnessed its immense power to benefit our lives. It is a development that ranks among the most impactful accomplishments ever made by the human race.

This history is told through a series of linked invention stories. Whether or not readers are previously familiar with the subject matter, these stories aim to be fascinating, informative, and directly relevant to our lives. Consider, for instance, the light bulb, invented by Thomas Edison. Every reader will be familiar with the device—but how many readers understand how the light bulb works? And of those, how many can explain how the light bulb was commercialized and what methods were employed to help ensure its market success? Furthermore, of those still fewer in number, how many can connect the dots, explaining how Edison's past led to the device's creation, and how the device itself led to the creation of power plants and the electric power grid? Will the new LED lighting render Edison's bulb obsolete in the near future? From the telegraph to the iPhone and for dozens of major discoveries and inventions in between, this book will help readers—regardless of their scientific backgrounds, whether extensive or nonexistent—*to connect the dots, fill their knowledge gaps, and gain total perspective about the continuous innovations that shaped our modern world.*

The journey of human beings' conquest of the electron is complex and long, with many intertwined epochal events and people involved. In this book, the series of linked invention stories are presented at a fast pace, in a chronologically consistent way, with a unique emphasis on insightfully delineating the *connectivity* of various inventions—that is, the cause and consequence of how one invention leads to another. For example, you likely know that Alexander Graham Bell is credited with inventing the telephone—but do you know why he was working on the device in the first place? As it happens, he was actually trying to improve the operation of Samuel Morse's telegraph; under pressure from his future father-in-law, he stumbled onto—or some might claim he stole—the novel idea of a telephone. History is full of cases like this. The technological innovation process is many-faceted. It is rare that inspiration or innovation take place in a vacuum; there are always specific scientific, marketplace, and personal agendas to consider!

Another emphasis for the book is to present *balanced case histories* of how major technical inventions lead to successful commercialization and ultimately impact the society. These cases clearly illustrate that no technical invention alone is adequate to usher in a new era. Besides the glamour of an invention, one also needs passion and courage for entrepreneurism, critical financial support, sound business management, and perfect market timing. An example of this is demonstrated in the series of cases that led from the invention of the transistor at Bell Labs to the birth of the microchip and personal computer, and the subsequent boom of Silicon Valley. Readers will see how, companies like Intel and Sony succeeded, while early pioneers such as Shockley Lab and Fairchild Semiconductor failed. These histories serve as valuable business cases for today's managers and entrepreneurs ready to build their own new technology-based businesses.

The continuous process of major breakthroughs in electronics technology has been like passing through a string of complex labyrinths, one after another. When looking back, many paths seem obvious; however, looking forward, when the future looms full of uncertainties and traps, the challenge to find the right path is never an easy one. Some of these maze-like hallways have proven to be dead ends, and entire lives and careers of brilliant researchers have wasted away in fruitless search for an exit. There have also been those who found one way out of their mazes, only to discover it was the wrong one, and they went astray on side paths that led to ultimate failure. Others succeeded in finding the right path—sometimes by pure chance—and still others followed them, broadening and leveling the way, allowing the next wave of explorers to cruise straight through in their wake. To the latecomers, the earlier mazes ceased to exist. For these people, their lot was to set their sights on the next mazes that lay further still beyond. Building on the successes of others, many trails would later merge to form an impressive, broader thoroughfare. Look, for instance, at the phone in your pocket: it represents the convergence of historical breakthroughs in wireless telephony, radio, television, GPS, and computing, as well as in component technologies such as the LCD and the incredibly versatile and powerful silicon chips! In this way, mankind has constantly pushed the frontier of technological development ever farther, and it continues to do so at an even more breathtaking pace.

The primary driving force behind all the progress is the creativity and dedication to entrepreneurism of individuals who successfully found or even stumbled onto the right path and marked the route for other people to follow. These are the *conquerors of the electron*, the people who made history and created a legacy that benefitted the entire human race. Some inventions required many years of arduous and persistent effort by the people involved, while for others, the results came from just a flash of inspiration. Even though major innovations often seem to be accompanied by a sense of inevitability, the roles played by these individuals were still crucial in shortening the timeline of discovery and influencing the direction of future development. What's more, rather than being deified as superheroes as many legends tend to do, the histories in this book depict these Prometheans as normal human beings, each with their own very personal and unique strengths, weakness, and idiosyncrasies. Some of them are true *geniuses* in certain specialty areas but total failures in others. Others are *visionaries*, but still more are *tinkerers* with a total lack of long-term thinking. There are egomaniacs or even *scoundrels* who prey on other people's accomplishments without regard for questions of right or morality. In fact, many of these traits can sometimes coexist in an individual. No matter how we may judge them, it is undeniable that each of these pioneers has contributed mightily to the conquest of the electron—a conquest that we have all benefitted from, every minute of our everyday lives. Their stories, therefore, should be known and appreciated. To enhance the reader's experience, we have also included numerous historical photos of these remarkable men in their prime, and made note of their accomplishments.

This book was originally written in Chinese and published in Taiwan in 2011 by Commonwealth Publication. It won the prestigious Golden Bookmark Prize as the best original Chinese language popular science book published in Taiwan, Hong Kong, and China from 2009 through 2011, out of over 400 entries. The content has been updated and significantly enriched for this first English language edition with important assistance and advice from Mr. Eric Brach, the co-writer.

Dr. Cheung would like to express his deep appreciation for a number of people who have assisted with the text, including Professor Frank Chang of UCLA; Nobel laureate Prof. Herb Kroemer of UC-Santa Barbara; Dr. Albert Yu, formerly senior vice president of Intel; and Dr. Milton Chang, a pioneer and a highly successful entrepreneur in photonics technology; Dr. Monte Khoshnevisan, a former colleague; and Elaine Cheung, his daughter. All of them have carefully read the manuscript and offered their insights and suggestions. He would also like to thank Prof. S. C. Lee, former president of National Taiwan University, for writing the introduction to the original Chinese version of the text, as well as Prof. K. Y. Chen, dean of engineering at National Tsinghua University; Prof. C. P. Wong, dean of engineering at Hong Kong Chinese University; Dr. Nim Cheung, chief executive officer of Hong Kong Applied Science and Technology Research Institute; and Dr. S. Y. Chiang, former executive vice president and co-chief operating officer of Taiwan Semiconductor Manufacturing Company, for their encouragement and endorsements. The author also acknowledges Mr. Craig Fennel for providing excellent graphics support. Finally, he would like to express his gratitude to his daughters Elaine, Vicky, and Yvonne for their support, and he remains deeply indebted to his wife, Jenny, for her constant support in helping him turn his vision into reality, as in all other endeavors.

I

AGE OF ELECTROMAGNETISM

1

THE KNOWLEDGE FOUNDATION

THE BEGINNING

Almost three thousand years ago, an unknown Greek recorded a mysterious phenomenon: when he rubbed a piece of cloth on a chunk of "elektron," or amber, the gemstone produced a mysterious, invisible force that could make one's hair stand on end and attract wheat chaff and the down of feathers. What this force was, exactly, was unknown to this forgotten Greek. Today, of course, we can readily recognize it as static electricity, but at the time, the phenomenon was unexplainable.

About four centuries later, in around 600 BCE, a farmer in the small town of Magnesia in Thessaly, Greece, discovered a rock that had the odd effect of invisibly attracting stone chips and fragments containing veins of iron. News of this rock—known as a lodestone—spread, and the words "magnet" and "magnetism" were coined, stemming from the name of that town.

As with many other natural phenomena during the early days of civilization, the basic principles governing static electricity and magnetism seemed incomprehensible. What's more, the Greek intellectual elite of the era generally advocated for philosophical reasoning to explain truth, often at the cost of ignoring a more methodological path to discoveries and invention. Thus, though political science and philosophy advanced swiftly in ancient times, understanding of natural sciences progressed only in fits and starts, their explorers cast to the jagged edges of the map of knowledge. The subsequent rise of the Christian Church in Europe only solidified this subjugation of research, as a wide swath of natural science found itself at odds with theology. For centuries, investigators of scientific truth were at risk of being declared heretics, and potentially even executed for their work. So for over two thousand years, European knowledge of electricity and magnetism never progressed much beyond that of the time of Ancient Greece.

Though Western understanding of electromagnetism remained limited for a long time, other cultures found valuable uses for electromagnetic phenomena. Independent of the Greeks, the ancient Chinese also discovered lodestones, and in 220 BCE, during the Warring States Period, Chinese sculptors discovered that magnetic objects possessed directional properties thanks to their attractive force. Surviving legends suggest that skilled craftsmen, using carving techniques borrowed from stonework in jade, fashioned lodestones into round-bottomed spoons that could freely revolve on a smooth and level "ground plane." It was discovered that, when the spoon came to a rest, its handle would always point toward the south.

Interestingly, this experiment has never been successfully reproduced in modern times. The weak magnetic forces from the earth are not strong enough to overcome the friction force between the spoon and plate to allow freedom of movement in all modern efforts to re-create the device, so there is still a shroud of mystery over the exact details of how these Chinese craftsmen were able to create the first working compass. But create it they did, and that compass, along

Ancient Chinese magnetic compass. Courtesy of Stan Sherer

with paper, printmaking, and gunpowder, became one of ancient China's "Four Great Inventions," the group of advances that had profound effects on the forward march of humanity.

As time progressed, the processes used to manufacture the compass—whatever they may have been—steadily advanced, and the compass's sensitivity and utility commensurately increased. In his "Dream Pool Essays," the Northern Song dynasty (960–1127 AD) polymath Shen Kuo described an advanced version of a compass made from an iron needle hanging horizontally in the air by a thread: rubbing the needle with a lodestone would cause the needle to acquire magnetic properties—to become "magnetized." Because a simple needle was far lighter than a carved spoon, this new design offered a clear improvement, and it was this new compass that was quickly adopted and applied across Asia, including for navigation in the open seas. Arab traders brought this compass technology to Europe around the beginning of the thirteenth century AD, whereupon it was widely adopted. Thus, two millennia after the principle was first discovered, magnetism was at last put into use on the European continent.

Though the Chinese merit due credit for creating and honing the compass, the fact remains that craftsmen of the ancient Chinese society engaged in little to no detailed study of the principles of magnetism. The Confucian attitude toward scholarship was not so different from the rational inference model prevalent in Europe during the time of Aristotle; the efforts and energies of the intelligentsia were dedicated not to understanding the natural principles of the physical world, but to the study of literature in preparation for the rigorous series of examinations required to enter employment with the state. The artisanal professions—under whose banner the creation of nearly everything fell—were not held in high esteem by that society, and they were often ignored by the best and the brightest. Thus, despite promising advances, the potential power of electricity and magnetism went largely ignored for nearly two thousand years.

THE SCIENTIFIC METHOD

Enormous change occurred in Europe from the fifteenth through the seventeenth centuries. The Renaissance swept the continent, followed by the Enlightenment; taken together, they were nothing short of a far-reaching, mind-liberating social revolution. For over a thousand years in the West, people's thought processes had been tightly fettered by the shackles of religious dogma. After the fall of Constantinople in 1453, however, intellectual visionaries gave rise to a movement to search for and revive those twin treasures of ancient Greek civilization: knowledge and truth. People pursued humanism and the laws of the natural world, and they began to take tentative steps toward the understanding of the principles governing various natural phenomena.

The most important development during the Renaissance that followed was a new methodology to approach these studies, known as "the scientific method." At that time, the term "science" did not exclusively mean the disciplines of mathematics, physics, and chemistry; rather, it referred to the process by which all new courses of study of the unknowns of the physical world should be carried out.

The scientific method called on researchers to follow a disciplined and objective process. First, the normal workings of the natural world had to be meticulously observed and a pattern established, after which point a "hypothetical" model was to be developed. Armed with this potential model, experiments were then performed to prove whether or not the proposed hypothesis was indeed valid.

This inductive method of experimenting to prove or disprove a claim quickly became the scientific standard. Universities, professional societies, and other similar institutions declared that validation through experimentation was necessary for proving the truth. It became a prerequisite when performing an experiment that all conditions and methods be open—that anyone could duplicate exactly what steps the experimenter had performed, and that anyone doing so would definitively arrive at the very same outcome, ensuring that all results were objective and impartial. At the time, this was a novel idea, completely different from the deductive rational inference method employed by Aristotle. Though no one could have foreseen it, from this point forward, Western scientific knowledge and technology would advance by leaps and bounds.

The experimenter who was to provide the basis for all future study of electricity and magnetism was the Englishman William Gilbert. Gilbert was one of the important figures of the Late Renaissance period. His family was wealthy and well connected; he was Queen Elizabeth I's Royal Physician. Gilbert not only argued for the importance of the scientific method, he also put it into practice when studying electrical and magnetic phenomena. He may have been history's first major experimental physicist, preceding even Galileo and the famous experiment atop the leaning tower of Pisa.

After over a decade of efforts and the expenditure of a substantial portion of his own fortune, in 1600, Gilbert published six huge volumes detailing the breadth of his research in a work titled "De Magnete," or "About the Magnet." These books contained all of his hypotheses and the results of his experiments.

William Gilbert. © Science Museum/Science & Society Picture Library

In this work, which is still available today, Gilbert elucidated numerous observations from his experiments, including the following conclusions:

Electricity and magnetism are two completely different phenomena.
The positive and negative poles of a magnet cannot be physically separated.
Electrical attraction disappears in water, but magnetic attraction survives.
Magnetic force disappears at elevated temperatures.

Gilbert also proposed two additional hypotheses that, at the time, he could not experimentally prove. The first was that the earth was an enormous magnet with its positive and negative centers at the South and North Poles. This, he claimed, was the reason that compasses always pointed directly along the north-south plane: they aligned themselves with the magnetic forces of the earth. His second hypothesis was that celestial bodies attracted and repelled each other magnetically. Though Gilbert firmly believed in these ideas, he did not know how to prove them or even test them experimentally. Instead, he simply elucidated them and left them for posterity to ponder. Of course many years later, scientists would prove that he was essentially correct in his first hypothesis about the earth's magnetic forces. However, he was wrong on the second hypothesis, as Newton proved that the interaction among the celestial bodies was due to gravitational forces, not magnetic force.

From a modern perspective, Gilbert's research could be labeled simplistic and crude. However, his work represented a major step forward in exploring electromagnetism, and it laid the groundwork for disciplined study of the subject in the future. Thanks to Gilbert's efforts, the human race was finally able to enter the maze of scientific exploration.

THE MAGIC OF STATIC ELECTRICITY

Despite the groundbreaking nature of the text, research into electricity and magnetism did not immediately blossom after Gilbert published "De Magnete." Like a seed, it would take a long gestation period before Gilbert's work broke through to the surface of scientific interest.

The first major step of the journey to expand the scope of Gilbert's ideas took place in 1663, when German pioneer Otto von Guericke invented a simple device. By rubbing a piece of cloth across a manually pedaled, rapidly revolving sulfur ball, he was able to build up a significant amount of static electrical charge on the surface of the sphere. Guericke did not have any application in mind for his invention; he created it purely out of curiosity. Historically, he was better known for his pioneering work in the invention of vacuum technology, which itself later would come to play a major role in the history of the development of electronics. Still, Guericke's charged sphere was a first: it established the possibility of generating and storing static electrical charge.

In 1729, more than sixty years after Guericke's invention of the static electricity generator, Englishman Stephen Gray became the first to discover that different classes of materials possessed different levels of electrical conductivity, or ability to conduct electric charge. He found that most metals could conduct electrical charges easily, whereas other materials, such as wood and cloth, could not. He also found that the earth itself—the ground—was electrically conductive. These findings inspired him to conduct a number of experiments, during which he discovered that he could transport electrical charges over substantial distances using metal conductors. For instance, using twisted wires and rods made of copper, Gray successfully transmitted static electricity generated by one Guericke Sphere to another sphere some 300 feet away.

Once he realized that electricity could be physically transferred, Gray conducted several well-known public demonstrations of the phenomenon of static electricity. The most famous of these was known as the "Flower Boy." In this experiment, Gray suspended a young boy in mid-air with non-conducting silk cords, electrically isolating his body from the conductive ground. He then transferred static electricity from a nearby Guericke Sphere to the boy's body using a metal wire. When the boy became statically charged, lightweight flower petals scattered on the ground would be induced by the electrical charge to rise and float around him, forming a mystical spectacle.

In 1733, Gray's friend, a young Frenchman named Charles du Fay, discovered that when producing static electricity with the Guericke Sphere, he could use different material combinations to produce two different kinds of charges: positive and negative. As with magnets, du Fay noted that electric charges of the same kind would be mutually repellant, while opposing kinds would be mutually attractive.

In 1745, the Dutch professor Pieter van Musschenbroek advanced the storage of electric charges beyond the simple Guericke Sphere by creating a specially designed glass bottle called a Leyden Jar, named after the Dutch university where he taught. The jar's structure was very simple. Metal foils were wrapped around both its outside and inside surfaces. These two foils were separated from each other, or insulated, by the glass in between. The interior metal foil could be charged from a Guericke Sphere, and when the two metal pieces—effectively, electrodes—were connected by a conductive connector, the jar would discharge its stored electricity in an instant, producing a spark and a shock. With the Leyden Jar, one could not only store electricity but also carry it elsewhere for discharge. Van Musschenbroek had effectively created an electric charge storage device: a capacitor.

The most famous story of a Leyden Jar discharge came just a year later, thanks to the work of the French monk Jean-Antoine Nollet. In 1746, Nollet performed an experiment on the grounds of his monastery. There, he arranged two hundred monks in a circle about a mile around. Every monk held a 25-foot brass pole in either hand, connecting them all in an unbroken chain. Nollet then connected the ends of the first and last brass pole to the electrodes of a specially designed, large capacity, fully charged Leyden Jar. In an instant, the unfortunate monks were all temporarily paralyzed when the electricity raced through their bodies. All of their muscles seized and clenched, as if they'd been struck by lightning. In this experiment, Nollet had intended to measure how quickly human bodies could conduct electricity. Not surprisingly, his conclusion was "with unlimited rapidity!"

Nollet's real aim was to find out how to use electricity to quickly transmit messages. He had intended to carry out additional experiments, but after just one trial, the now-wiser monks knew better and refused to cooperate. With nowhere else to turn, Nollet petitioned the French King, Louis XV, at Versailles. Sensing the potential but understanding the reticence of the general public, the king granted Nollet use of 180 Royal Guards—unlucky for them—in his next experiments. Nollet's name—at least in some circles—became, thereafter, quite well known.

Another major contributor to the study of static electricity was one of America's Founding Fathers, Benjamin Franklin. Franklin is said to have been inspired to study electrical phenomena by observing the discharge of two electrically charged spheres. When the objects got close enough to each other, sparks and arcs flew across the gap—sparks and arcs that, to Franklin, looked just like a miniaturized version of lightning in the sky. In a flash of inspiration, Franklin, the ever inquisitive and creative person, decided to do a bold and clever experiment to prove that lightning is, in fact, caused by the discharge of static electricity stored in cloud layers.

In 1752, on a day of thunderstorms, Franklin flew a kite into the sky. This was no ordinary kite, however: the kite's string had been substituted with a length of thin copper wire. On the end of the long wire, Franklin fastened a heavy set of brass keys, which were then placed inside a Leyden Jar. When the lightning flashed across the sky and hit the kite, the resultant electricity was transmitted from the sky via the copper wire to the brass keys in the jar, filling the previously uncharged Leyden Jar with electricity from heaven.

Historically, this is a well-known event. However, it has never been authenticated. A few years after it was supposed to have taken place, a Danish man attempting to carry out the same experiment was killed after being struck by the lightning. What *is* known for certain about Franklin and lightning, though, is that Franklin invented the lightning rod. In Franklin's time, most of the frames of buildings were constructed with wood, and when a house or building was struck by lightning, there was a likelihood that the edifice would catch fire. Franklin, however, had an idea. To safeguard a structure, he proposed attach-

Franklin and son flying kite in a thunderstorm with lightning. Sheila Terry/Science Source

ing a pointed metal rod to its roof and running a copper wire from the bottom of the rod to the ground, where it could be buried. This arrangement would effectively short-circuit dangerous lightning strikes by diverting the electric charge of lightning from the building into the earth.

The lightning rod marked the first time that mankind applied knowledge of electricity to solve a practical problem. In Europe, the frequency of damaging lightning strikes, especially to churches, was so great that, for years, people believed that these bolts were special messages sent from God. They did not realize that the tall steeples with metal crosses atop them made the buildings ready targets for lightning to strike! At first, European churches refused to install the lightning rods, thinking that they would interfere with Divine Will. But when people saw the rods' efficacy, they were installed quickly, quietly, and without any fuss. Franklin's device proved to be of great value—but had he invented the same device just a few centuries earlier, he might have been branded a heretic and burned at the stake!

While Nollet, Franklin, and others were making important advances to the qualitative understanding of electricity, important quantitative principles were also established by French engineer Charles-Augustin de Coulomb. Coulomb was a member of France's Royal Institute of Science and was commissioned to find ways to improve the accuracy of the compass for the French Navy. The compass technology used up to that time was very similar to the Chinese design with the suspended, magnetized needle. Coulomb observed that as the magnetic needle revolved, the string that held it aloft began to twist. The more the string became twisted, the more it began to exert counterforce against the compass needle as the string tried to unwind itself. This counterforce—torsion—naturally affected the compass's accuracy.

Coulomb became fascinated by this, and he dedicated himself to conducting a series of experiments designed to precisely calibrate the torsion force of thin wires made from different materials. Along the way, he also realized that he could use this method to precisely measure the magnitude of interacting magnetic or electrostatic forces.

In 1789, Coulomb completed his experiments and published his findings. He proved that the magnitude of interacting electrostatic forces between two charged bodies is proportional to the product of the charges on each, and that it is inversely proportional to the square of the distance between the bodies. This equation became known as Coulomb's Law, and it marked the first quantitative success in understanding electrical science.

Nollet's monks, Coulomb's compass, and Franklin's kite . . . such stories filled the public with curiosity and awe about electricity. It even became fashionable for people to get a quick "shot" of electricity from Leyden Jars kept behind the bar in pubs and taverns; such an experience became known as the "electric kiss." What really captivated people's imagination, though, was the discovery of electricity residing within the bodies of living creatures, a natural phenomenon that led to one of the most important advances in electrical history.

THE BATTERY

In 1757, a French botanist discovered a new type of catfish while traveling in South America. When he touched it, he immediately felt a shocking sensation, much like a kiss from a Leyden Jar. Might this catfish's body, he wondered, somehow contain electricity?

In 1772, this discovery was matched by that of a British colonial official in India who came across a type of eel whose body appeared to be totally charged with electricity. In fact, this eel could even produce sparks that flew across its skin! These discoveries led some thinkers to deduce that there must be electricity residing in some animals'—perhaps all animals'—bodies. This idea led them to further surmise that "animal electricity" must somehow be related to the existence of life itself. Though these notions did not prove to be true exactly as they were originally conceived, this study of "animal electricity" unexpectedly ushered in a new era of knowledge of electromagnetism.

When people discovered that some living creatures appeared capable of not only conducting electricity via their bodies, as Gray and Nollet had shown, but also storing and employing the charge, it triggered a wave of research into animal electricity phenomena. Among the leading researchers in this field was Luigi Galvani, a professor of anatomy at the University of Bologna.

Whenever Galvani dissected a frog, he would lay it on its back atop a copper pan and pin it down with a metal clip made of zinc. Over time, he noticed that each time his clip brushed against the body of a freshly incised frog, its thigh muscles would twitch, just like the muscles of somebody getting an electric shock. Galvani surmised that fluid in the frog's muscles stored animal electricity, and he published this theory.

Galvani's controversial findings became, within the scientific community, a hot topic of debate. Among the opponents of Galvani's theories were an old friend and colleague named Alessandro Volta. Volta, a professor of chemistry at the nearby University of Pavia, was well known and respected for his discovery of CH_4, or methane gas. Volta's opinion was that the twitching of the frog's muscles was not due to the stored electricity of the frog itself, but rather a product of Galvani's using two different types of metal to touch it. The shock, Volta claimed, was created by the interaction between the two different metal contacts.

Animosity grew between the two as each publicly denounced the other's hypotheses. The dispute grew more and more intense over time, and eventually, the two friends fell out completely. Galvani later showed that even when he used just one kind of metal to touch the frog, its leg muscles would still twitch. Volta would not accept defeat, though. Instead, he redoubled his efforts to prove that he was right and that it was the metal contacts, not the frog that created the observed phenomenon.

In March of 1800, Volta proved that he could produce continuous electric current by connecting different metals—and he did not require the muscles of a frog. He announced to a stunned world his monumental invention: the Voltaic Pile, a stack of thin, alternating zinc and silver discs sandwiched between layers of brine-soaked cardboard or felt. This "artificial electric organ," as he termed it, could produce and store electricity. When the top zinc disc was connected to the bottom silver one by a copper ribbon, a steady stream of electric current began to flow along the copper strand. Galvani died one year before Volta's discovery and so was never able to witness the demonstration and continue the debate.

This invention represented a far stride past the capabilities of a Leyden Jar. The electricity that could be stored in a Leyden Jar was quite limited, and there was no way to temper its discharge—as Nollet's monks had learned long ago, the electrical buildup in a Leyden Jar discharged itself completely in one quick, exhaustive burst. But the Voltaic Pile—later renamed a "battery" by Benjamin Franklin —provided a continuous and controllable source of electricity. Furthermore, the total amount of electric charge generated from Volta's battery was far greater than that of a Leyden Jar, and its continuous flow of electric current was stable and could be more accurately gauged and reproduced.

In keeping with the spirit of the scientific method of research, Volta fully disclosed the detailed design of the battery in a published paper. Scholars all across Europe soon began to duplicate and even improve upon Volta's device. Volta's design was simple, and the materials he used were inexpensive, especially when he showed that the silver plates could be replaced by copper plates. What's more, his design was *scalable.* If a higher potential energy of electric charge—later known as "voltage"—was required by the user, more batteries could be stacked on top of one another. For a larger current—that is, a higher flow rate of the electric charge—the surface area of each disc could be enlarged or many batteries could be linked in parallel. With the invention of the battery, mankind had finally found a reliable, consistent, and continuous source of electricity.

Volta was basked in glory after his invention of the battery. No less than Napoleon Bonaparte, then the lord and ruler of the northern part of Italy, made a special request to ask Volta to explain to him his invention. Volta demonstrated that using the battery as a source of current, he could heat an iron wire to red-hot, or he could produce gas bubbles when dipping that wire in water. Napoleon, impressed, invited Volta to take up a government position in Paris and, with it, a generous stipend. Volta accepted, and from that point on, he never did

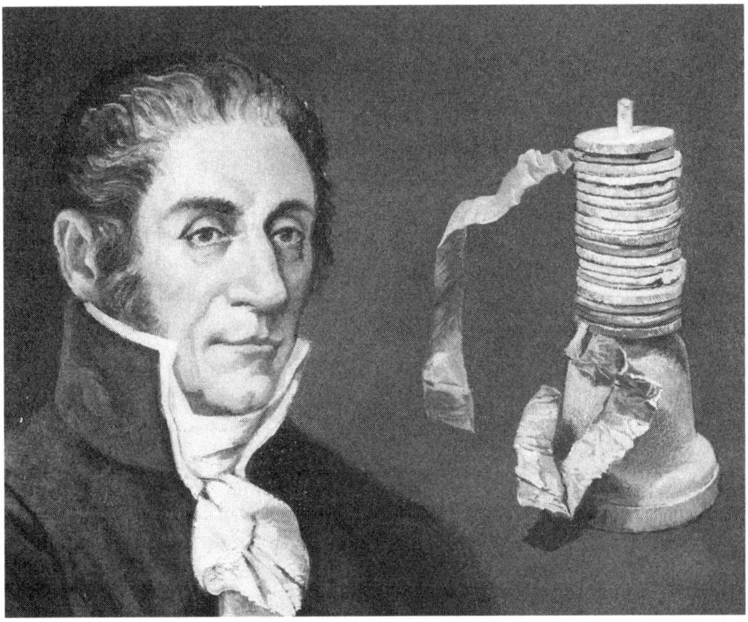

Alessandro Volta and his "Pile." UIG via Getty Images

any further serious research. However, his impressive work on the battery more than cemented his impact and his legend. Even today, Volta is remembered well by his countrymen: alongside the beautiful Lake Como stands the majestic Volta Temple, facing the snow-capped Alps. Among all the scientists and engineers in history, very few can match the physical tribute Volta has received.

LINKING ELECTRICITY AND MAGNETISM

Volta-type batteries became commercially available within two years of their unveiling, thanks to Volta's willingness to make public all information related to his creation. Subsequent inventors and researchers made refinements to Volta's design, and battery performance improved. Because batteries provided a stable and relatively long-lasting source of electricity, the reliability and repeatability of electrical experiments also increased. This led to a rapid expansion of the research community and the buildup of knowledge about electricity, including the birth of a new branch of science: electrochemistry. The most influential researcher in this new branch of science was an English chemist named Humphry Davy, who worked at the Royal Institution.

Founded in the mid-eighteenth century, the Royal Institution was a non-profit private organization in Great Britain. Its mission was to promote scientific advancement and popularize scientific education in society. The Institution provided financial support to basic research in select fields, and it also sponsored seminars and lectures for the dissemination and exchange of scientific knowledge to the general public.

The Royal Institution had a dedicated building in London that contained laboratories, meeting rooms, and a public lecture hall, as well as a small set of living quarters in the attic for a few employees. Davy, who was the director of the Institution and its chief scientist, built an enormous battery in the Royal Institution's basement consisting of 250 linked voltaic piles. He used this extremely powerful battery to perform unique experiments that were beyond the capability of most other researchers.

Amidst all kinds of new experiments using the battery, the earliest to arouse attention involved electrolysis and electroplating. It was discovered that passing electrical current through certain types of liquid solutions led to chemical reactions. Substances could be coated with aqueous ions of metals, and dissolved solids could be precipitated out of their liquid states. Davy successfully electrolyzed many complex chemical solutions and analyzed the composition of their elements. In so doing, he became the first man to isolate and identify many new elemental materials, including potassium, sodium, calcium, strontium, barium, magnesium and iodine. The discovery of these elements would probably have been delayed for many years without the tools provided by the study of electricity and the invention of the battery.

Humphry Davy. Sheila Terry/Science Source

In 1807, Davy conducted another important experiment, very different from his work analyzing aqueous solutions. In it, he sharpened two electrically conductive carbon rods and connected them to opposite electrodes of a high-voltage battery. He then began to reposition the twin rods ever closer until suddenly, an electrical current flowed across the gap and through the air, a brilliant white arc flashing between the rods' tips! The principle was very similar to Benjamin Franklin's observation of sparks shooting between two charged spheres back in 1752, but this marked the first time electricity had ever been used deliberately by mankind to create sustained light. It was a historic moment that would, years later, open the door to the electronic age.

With Volta's invention of the battery and Davy's advances in England, scientists in continental Europe increasingly became attracted to the study of electricity. Scholars who had originally specialized in Newtonian mechanics or mathematics now began to dip their toes into the charged water. This enlarged pool of scientific talents greatly accelerated the progress of understanding of electricity and magnetism.

In 1820, when the Danish physicist Hans Oersted was teaching a class at the University of Copenhagen, he happened to notice an unexpected phenomenon. While he demonstrated passing a current through an electrical wire, he inadvertently saw the needle of a compass on his podium beginning to quiver! He turned off the current and the compass needle immediately returned to its original state. Could it be, Oersted wondered, that electricity and magnetism were in some way related? That certainly seemed to be the case. But hadn't Gilbert stated long ago that electricity and magnetism were two different phenomena? Oersted could not explain what had happened, but he was convinced that the observation was real and the implication was important. Oersted soon published his startling observation, and it quickly became the talk of the European academic establishment.

Only two months after the results of Oersted's accidental discovery were published, the French mathematician André-Marie Ampère performed a still more detailed experiment. He placed two electrical wires close together and ran currents through both of them. When the currents in the wires ran in the same direction, the two wires repelled each other. But when the currents in the wires flowed in opposite directions from one another, the two wires attracted each other. Ampère deduced that these forces of mutual attraction and repulsion came from the magnetic fields created by the electric currents. Furthermore, he claimed, it was this current-induced magnetic field that made Oersted's compass needle rotate. Ampère, an exceptionally brilliant mathematician with tremendous physical insight, formulated the first complex mathematical model to explain the relationship between electric current and magnetism. Gilbert, the father of the field, had been proven wrong.

In addition to Ampère's breakthrough, there were many valuable practical physical phenomena observed during this time as well. The most significant was the discovery of the first practical electromagnet in 1825 by an Englishman named William Sturgeon. He took a small rod of non-magnetic "soft" iron material and wrapped it with electrical wire. He then bent the rod into a horseshoe shape. Amazingly, when electrical current passed through the wire, the soft iron core was turned into a magnet, with the two ends of the horseshoe behaving as the two opposing magnetic poles. The induced magnetic force disappeared as soon as the electrical current stopped flowing. This electromagnetic phenomenon could not be explained by scientists of the time, yet it found many important applications.

FARADAY, THE GRAND MASTER

With the invention of the battery and the beginning of the unification of electrical and magnetic phenomena, scientists and mathematicians around the globe, such as Germany's George Ohm, Carl Gauss, and Wilhelm Weber, and an American, Joseph Henry, all made important contributions to the understanding of electromagnetism. Still, however, factual understanding of the subject was limited to a few specific areas, and it would remain so until the emergence of two extraordinary geniuses who greatly elevated the breadth and depth of understanding of electromagnetism. The first among them was an Englishman named Michael Faraday.

Faraday was one of four children born into a desperately poor family in London. His father was a chronically ill blacksmith without a steady income, and all the Faradays lived in a state of near starvation. What's more, schooling at that time was neither compulsory nor free, so apart from what learning his mother could impart at home, there was no opportunity for the boy to receive a formal education.

At the age of fourteen, Faraday began a job as an apprentice in a bookbinder's shop. Although afflicted with a stutter, he had an extremely inquisitive mind and a thirst for knowledge, so, whenever he had spare time, he would read the books that he was binding. He liked the Encyclopedia Britannica best, especially its descriptions of the cutting-edge sciences of electricity, magnetism, and chemistry.

In its efforts to promote scientific knowledge, the Royal Institution frequently held seminars and lectures that were open to the public. On February 29, 1812, young Michael Faraday received a ticket from a friend to attend a lecture by Royal Institution director Humphry Davy. As Davy lectured, Faraday took thorough notes, attending to Davy's every word. This generous friend

Michael Faraday. SPL/Science Source

noted Faraday's intense interest and secured him tickets every time Davy gave an open lecture. The more he attended, the more fascinated Faraday became with the burgeoning fields of electricity and chemistry. Finally, Faraday mustered his courage to approach Davy and ask for employment as an entry-level assistant in his laboratory. In order to flatter the man, Faraday called upon his modest skills and presented Davy with a gift: a handsomely bound book containing all the notes Faraday had taken at Davy's lectures. Still, Davy—at least at the outset—was forced to turn the boy down.

It's funny the way fate works—oftentimes, an event occurs that could just as easily have not. It's when these happenstances greatly alter the course of the future that observers are left scratching their heads. What if Faraday's friend had not supplied him with tickets? Then Faraday would never have become enamored of Davy and his lectures. What if Faraday had worked as an apprentice in some trade other than bookbinding? Then he may never have had the opportunity to be exposed to the knowledge contained in Encyclopedia Britannica, nor would he have had the chance to present Davy with such a memorable gift. And what if a seasoned experimental assistant in Davy's office didn't quit just days after Faraday approached Davy and tried to win his approval? Then Davy would likely have never changed his mind and offered a position to the young Faraday, and history would have been forever robbed of one of its greatest scientific minds.

Although he was just twenty-one years old and not formally schooled when he became Davy's assistant, Faraday quickly proved himself to be exceptionally proficient. He was skillful in hands-on experiments and in short order became indispensable to Davy. In addition, though Faraday had never been trained in any aspect of higher mathematics, he had uncanny intuition and a superhuman ability to visualize abstract objects, concepts, and shapes.

Even though Davy recognized Faraday's talent and found his service very useful, deep in his heart, he had little respect for this socially lower-class bookbinder. Davy and his wife took Faraday along during an extended tour of mainland Europe, but during their travels, they treated him almost like a servant. Nevertheless, this trip afforded the young man a chance to meet the top scientists at the time, greats like Ampère and the aged but still-kicking Volta. These encounters added greatly to Faraday's personal experience and self-confidence.

Upon returning from continental Europe, Faraday began taking on more responsibilities in the laboratory, which freed Davy to accept more lucrative lecturing tours. While Davy traveled, the young Faraday began to carry out experiments based on his own ideas. In 1821, just one year after Oersted and Ampère discovered the linkage between electricity and magnetism, Faraday performed what became an experiment of historic importance.

Faraday fixed a magnet inside a wide-mouthed glass beaker and half-filled the beaker with electrically conductive mercury. Then, taking a thin metal rod, he dipped one end into the mercury and loosely attached the other end to a metal hook. When he passed electrical current through the mercury to the metal rod, something amazing happened. Like magic, the metal rod began to slowly revolve around the magnet. When Faraday switched the direction of the electrical current, the rod revolved in the opposite direction. The physical principle governing this observation is an extension of Ampère's demonstration

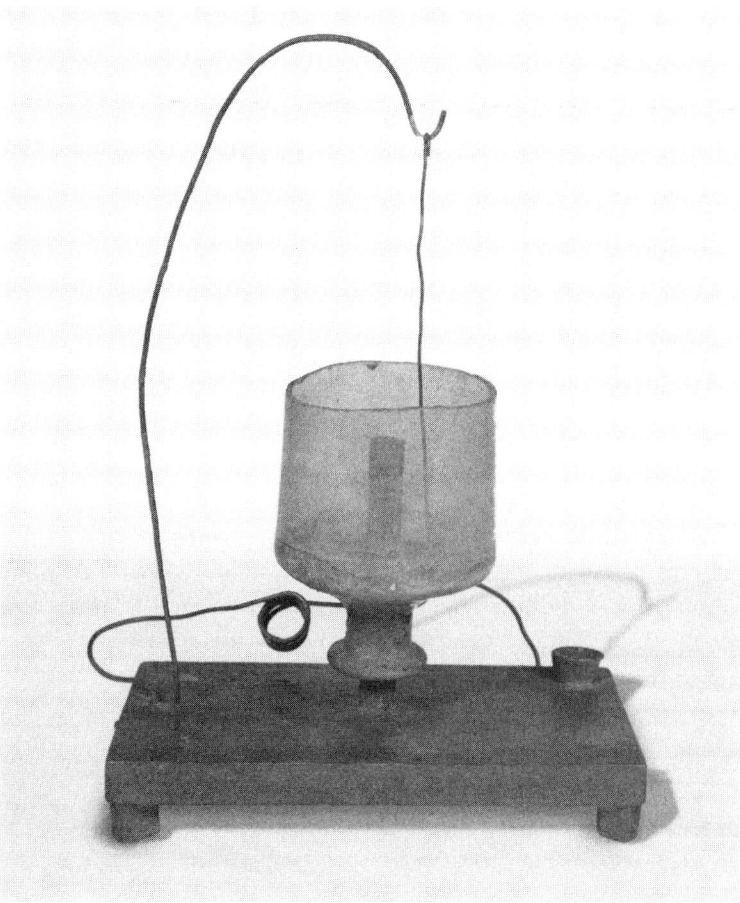

Faraday's motor experiment (replica). Courtesy of Spark Museum

of attracting and repelling wires. In Faraday's case, the magnetic force created by the electrical current flowing in the rod interacted with the magnetic force of the magnet situated in the center of the cup, causing the two to push each other away. Coupled with the limited range of motion afforded by the affixed, hanging wire, this repelling force caused the rod to revolve around the magnet. This simple yet elegant experiment demonstrated for the first time that electrical energy could be directly converted into kinetic energy, the energy of motion. Faraday had created the foundation for the first electrical motor!

Faraday published the results of this experiment to great fanfare. However, he attracted quite a bit of envy amidst all the applause, not least of which came from his mentor and idol, Humphry Davy. Davy, it seems, was particularly upset because Faraday had not listed him as a co-author of the groundbreaking paper. As Faraday saw it, Davy played no role whatsoever in this experiment and thus made no contribution to it; therefore, he was right to be excluded. Davy, on the other hand, felt that it was his lab and Faraday his pupil and subordinate, and thus he himself should receive some of the recognition.

This issue grew to become a very sore point between the two men that never healed. Though Faraday continued to work under Davy, he constantly felt as if he were treading on thin ice. Davy never again assigned Faraday to work on any experiment involving electrical and kinetic energy. To avoid friction in the workplace, Faraday himself intentionally avoided the field for the rest of his tenure under Davy, instead focusing his research on electrochemistry exclusively. Interestingly, this may have proved beneficial for not only personal reasons, but for professional ones as well—in devoting himself to electrochemistry, Faraday eventually emerged as the great master in this field. He established the principles of applied electrochemistry for batteries, electrolysis, and electroplating. He also discovered the ring structure of benzene, which became of great importance in organic chemistry, and the material itself has numerous applications. It is through his work in electrochemistry that Faraday also solved the great debate

Faraday lecturing at Royal Institution. SPL/Science Source

of a generation before: the debate between Volta and Galvani of the true source of electricity.

By the time Volta unveiled his battery to the world, Galvani was already dead and much of his work on animal electricity was pushed to the background. However, Volta's findings had never logically proven that Galvani's hypotheses about tissue-borne electrical charges were incorrect. Volta only showed, or thought he showed, that two different metals could produce an electric current via a chemical interface. But Faraday's later studies of electrochemistry determined that the source of the continuous electrical current in a Voltaic Pile was not due to the metal plates alone. Rather, it was the result of chemical reactions between the metal electrodes and the electrolytes—the ionized salts—contained in the brine that soaked the cardboard or felt sandwiched between those metal plates. Thus, though Volta had emerged the clear winner in his very public debate, Faraday eventually showed that the conclusions reached by this groundbreaking genius had not actually been completely correct. Moreover, many years later, Galvani's reputation was also salvaged when nerve endings were found to operate via neuroelectrical impulses and even later, electromagnetic brain waves were discovered. So ironically, how should the winner or loser ultimately be judged in the case?

When Humphry Davy passed away in 1829, Michael Faraday became the new director of the Royal Institution. His appointment was backed by popular demand, and Faraday, once the student, now the master, began to present his own keynote lectures to the public. In addition, as he no longer had to face the disapproval of his former mentor, Faraday slowly resumed his studies of the intersection between electricity, magnetism, and kinetic energy.

Faraday, ever a believer in the intrinsic symmetry of nature, focused his new efforts on demonstrating that a change in magnetic field could generate electrical current. He had already shown with his motor experiment how current could be converted into motion in the presence of magnetism; the inverse, he felt, must also be possible. However, the technical challenge proved difficult, and for a long time, his experiments failed. At last, in 1831, when Faraday was in the process of inserting a magnet into a coil of wire, he noticed that an indicating needle on a meter measuring the current in the coil appeared to move ever so slightly before returning to its original position. Almost by accident but thanks to his keen observation, Faraday discovered that when a magnet and a coil of wire move with respect to one another, this motion produces an electric current in the wire—in other words, that kinetic energy of motion could be converted into electric energy. It was a eureka moment as the principle of the electrical generator was founded, even though it would take another forty years of technological advancement before motors and generators could be made practical.

In addition to developing the principles of electrical motors and generators, Faraday also discovered many other important electrical and magnetic phenomena, such as electrical induction. The induction principle is used extensively in many important daily applications, ranging from power transformers and countertop cooking surfaces to wireless chargers and contactless readouts for debit cards and room keys.

Faraday's impact was far-reaching in part because his intuition was so extraordinary. For instance, when in his later years he found himself recovering from illness in Switzerland with little to do but lie in bed and observe, Faraday intuited that the sunlight streaming through his window might be a form of oscillating electrical and magnetic forces. To many people, this sounded like a bizarre idea, but, using a simple electromagnet, a piece of quartz, and a polarizer film, he was able to experimentally prove that he was correct. In so doing, Faraday discovered the magneto-optical effect—the ability of magnetic forces to change the polarization of light waves—and for the first time in history linked electromagnetism with light.

Faraday devoted his entire career to the study of electromagnetism. Even as the director and one of the world's foremost scientists, he and his wife lived humbly in the attic of the Royal Institution—the same place Faraday had lived since he was a poor bookbinder-cum-lab assistant. His motto, "Work, complete, publish," embodied the spirit of a true scientist in searching for and sharing truth in nature. He never once applied for a patent despite the potential financial windfalls attendant to his work; instead, he chose to leave all the knowledge from his long and highly productive career for the benefit of the entire human race. Faraday meticulously recorded every experiment he conducted—close to six hundred in total—even though most were failures and did not produce the results that he initially sought. Today, all documents related to Faraday's work are preserved in the Faraday Museum annex of the Royal Institution.

Despite his genius, Faraday died with some work unfinished, leaving for the future a number of important riddles to solve. Among them was the concept of electrical and magnetic "fields." Faraday observed that when two magnets were placed near one another, they would be subject to mutually attracting and mutually repelling forces. As the distance between these two objects widened, these forces would correspondingly weaken, analogous to Coulomb's Law for static electrical forces. However, the question Faraday wrestled with, and never solved, was how these forces were transmitted through space without any physical contact.

Faraday liked to place a magnet under a piece of paper and then sprinkle a few iron filings on top. He would observe the filings' ovular distribution, highlighted by stark lines like stretched elastic bands or springs. He called these directional markers "lines of force." Faraday noted how the distribution of iron filings would change when a second magnet was placed nearby. Each magnet, he saw, had its own "sphere of influence" in which it acted on any other magnetic substance. Faraday named the distribution of such magnetic forces a "magnetic field." He extrapolated that within such a field, magnetic forces existed at every point in space, each with its own levels of intensity and directionality. Therefore, when two magnets were placed in proximity to one another, interaction should occur at the places where their magnetic fields overlapped. The resulting magnetic force at each point would be an amalgamation of the two forces.

Faraday's intuition told him that the forces that remotely acted on magnets were derived from these interactive fields and their lines of force. But Faraday, despite his genius, did not possess the advanced mathematical ability necessary to carry this idea beyond its qualitative, conceptual stage. That job would eventually be tasked to a man four decades his junior: James Clerk Maxwell.

MAXWELL, THE PEERLESS GENIUS

James Clerk Maxwell was born in Edinburgh, Scotland in 1831, the same year Faraday created his generator. Unlike Faraday, Maxwell's family came from means, and his father, a prominent lawyer, had no trouble supporting his son with advanced formal education. From early on, Maxwell showed genius in mathematics, and he was particularly outstanding at applying mathematical principles as a language to concisely and precisely explain natural phenomena. At the young age of twenty-eight, he was appointed to an assistant professorship at prestigious Cambridge University, his alma mater.

While Maxwell was still a student at Cambridge University, a celebrated fellow Scottish scientist and family friend, Professor William Thomson (later to be known as Lord Kelvin, a key figure in the development of thermodynamics, electromagnetism and its practical application in telecommunication), became extremely impressed by his talent and encouraged him to study the works of Faraday. Perhaps, suggested the professor, Maxwell might be able to use his talent in mathematics to interpret and unify the many phenomena that Faraday had demonstrated in his experiments.

Maxwell's exceptional mathematical ability became evident when he participated in a contest to deduce the characteristics of the rings around Saturn. It was a question that had stumped experts in the fields of physics and astronomy for years. (Telescopes of the time were not powerful enough to discern the details of the ring.) Though Maxwell knew little of the heavens, he was a math

James Clerk Maxwell as a student at Trinity College. Courtesy of Master and Fellows of Trinity College, Cambridge

genius, the kind of wizard who possessed the numerical skills that Faraday had only wished to have to accompany his insight and vision.

Using only mathematical analysis, Maxwell reached the conclusion that the rings of Saturn could not be monolithic pieces of solid material or suspended liquid. Instead, he deduced, they could only be floating bands of loosely arranged solids of varying sizes revolving around the planet. His analysis was so convincing that he won the contest and collected the cash prize of 139 sterling pounds. When the Voyager unmanned spacecraft flew past Saturn over a century later, the photos sent back to Earth proved Maxwell correct.

Before Maxwell devoted himself totally to the study of electromagnetics, he made major contributions to a number of other fields at the intersection of mathematics and physics, including astronomy, control theory, statistical mechanics, thermodynamics, and the physics of mixing colors. He even printed the world's first color photograph. But in 1862, not long after the fledgling outset of his professorial career, Britain's Royal Society, with the recommendation of Prof. Thomson, commissioned the thirty-one-year-old Maxwell to attempt to unify the basic units of electromagnetism. Maxwell's first step was to call on the master: the now-seventy-year-old Faraday. Faraday shared all his thoughts with the young man, and he hoped that someday the young math genius would be able to put all the pieces in perfect order.

Maxwell combined Faraday's insights with the ideas of Ampère, Gauss, Lord Kelvin, and numerous other scientists. He pondered the complexities of the relationship between electrical and magnetic fields, and using principles of Newtonian mechanics, fluid mechanics, and thermal dynamics, he tried to mathematically model the behavior of all electromagnetic phenomena. After eleven laborious years, Maxwell finally reached the finish line, and in 1873 he published his classic work: "A Treatise on Electricity and Magnetism." Using four seemingly simple differential equations (originally twenty, but later condensed to four by British physicist Oliver Heaviside), Maxwell was able to embody all the electrical and magnetic principles under the sun!

This quantitative work was the pinnacle of nineteenth-century scientific development, a transcendent discovery for humans seeking to unearth the truths of the natural world. It would be only a slight exaggeration to say that Maxwell's equations made some otherwise agnostic scientists believe in a Creator who had devised these four perfect principles, allowing the various properties of electricity and magnetism to henceforth operate under simple laws.

Maxwell's equations could explain qualitatively and quantitatively all the electrical and magnetic phenomena that had been observed to that time without any exceptions or contradictions. Impressive though that was, still more astounding was that these equations also predicted the existence of still-unidentified electromagnetic waves, which were expected to result from the oscillation of electric and magnetic fields. Furthermore, the travelling speed of these yet-unseen electromagnetic waves could be derived using several well-established universal constants.

$$\nabla \cdot \vec{E} = \frac{\rho}{\varepsilon_0}$$

$$\nabla \times \vec{E} = -\frac{\partial \vec{B}}{\partial t}$$

$$\nabla \cdot \vec{B} = 0$$

$$\nabla \times \vec{B} = \mu_0 \vec{J} + \mu_0 \varepsilon_0 \frac{\partial \vec{E}}{\partial t}$$

Maxwell's Equations. Derek Cheung

To everyone's astonishment, the theoretically predicted speed of travel of these waves was approximately 300,000 kilometers per second—within experimental error, identical to the speed of light! Though Faraday was no longer alive to see it happen, the observation he made on his sickbed—that light was a type of oscillating electromagnetic wave—was therefore proven mathematically by Maxwell.

The publication of Maxwell's equations created a storm of sensation within the scientific community. Despite the rigor of his mathematics, many naysayers remained dubious of his conclusions, particularly the idea that there were other electromagnetic waves beyond the realm of visible light. Scientists had already determined that the sensory cells of the human retina reacted only to the specific wavelengths of visible light, allowing people to see. But according to Maxwell's equations, electromagnetic waves of all wavelengths could exist, including at wavelengths much longer or shorter than what humans could perceive. So what kind of "light," his detractors cawed, could these other electromagnetic waves be?

After Maxwell published his work, a group of mostly British scientists known as the Maxwellians began to conduct a deliberate search for these invisible electromagnetic waves. However, despite fifteen years of intense experimentation, they came up empty-handed. It wasn't until the publication of the work of a young and brilliant German physicist named Heinrich Hertz in 1888 that Maxwell's proofs were finally universally accepted.

The test performed by Hertz to verify Maxwell's prediction was to generate and detect electromagnetic waves that were outside the realm of visibility to the naked eye. The wavelength of the electromagnetic waves Hertz generated was about one meter long, approximately one million times longer than the wavelength of any visible light. Still, the properties of Hertz's waves were completely consistent with Maxwell's mathematical predictions.

So what was this mysterious "light" outside of the visible spectrum that Hertz had created? It turned out to be something we are all very familiar with today: the radio wave. Hertz found that the properties of radio waves and light are basically the same: both obey the Maxwell equations, and they differ from one another only in their wavelengths. After the discovery of the radio wave, a whole series of new electromagnetic waves was discovered, including Gamma rays, X-rays, and microwaves. Some previously unexplained physical phenomena discovered before Maxwell's theory, such as infrared and ultraviolet radiation, were later also proven to be electromagnetic waves. It was ironic that the Maxwellians spent fifteen years of fruitless efforts to detect a mysterious invisible electromagnetic wave, yet once Hertz made his breakthrough, people realized that Maxwellian waves were in fact everywhere! Sadly, Hertz's career was itself cut short not long after he published his findings—he died at the young age of thirty-six when he suffered septic shock

Heinrich Hertz. Science Source

after an operation to remove a tumor. This abruptly ended the career of a brilliant scientist with exceptional ability in both theory and experimentation. Similarly, Maxwell himself never had the opportunity to see his theory validated by Hertz—he died of stomach cancer in 1879 at age forty-eight.

Scientific understanding of electricity and magnetism began with Gilbert's work in 1600. From there, it took nearly three hundred years for that study to reach its pinnacle with Maxwell's equations and their full validation via Hertz's discovery of radio waves. This was quite a long gestation period, but recall that these founders were forced to construct their entire knowledge base from almost nothing. Nonetheless, by the mid-nineteenth century, enough knowledge had been accumulated that important applications began to emerge, which in turn rapidly accelerated further technical and scientific progress.

The scientific knowledge foundation built by these forerunners was extremely robust. Even today, the work of those seemingly ancient scientists continues to shape our world. The fundamental principle of the battery remains more or less unchanged from Volta's time, and we still use Maxwell's equations extensively, such as in designing the most efficient antennae configurations for cell phones or in shaping stealth airplanes that can evade radar detection.

For years, electrical currents and electromagnetic waves were mysterious and complex phenomena, dangerously close to falling into the realm of pseudoscience. But two things saved them: the open and objective scientific method, and the research done by some of history's greatest minds. Many years later, a genius in his own right named Albert Einstein adorned his office walls with the portraits of three of his most admired scientists: Isaac Newton, Michael Faraday, and James Clerk Maxwell!

Electromagnetic Waves

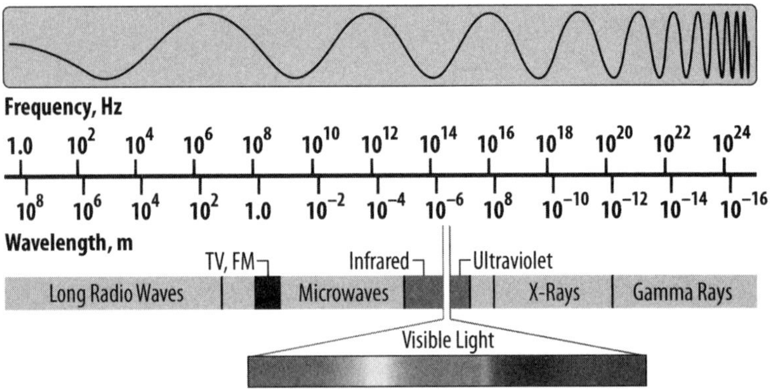

Electromagnetic Wave Spectrum. Derek Cheung

The electromagntic wave is a periodic oscillation of peaks and valleys of electric and magnetic fields which are perpendicular to each other and to the travelling direction of the wave. The physical length between each repeating period is known as wavelength, and it is expressed in meters. All electromagnetic waves travel through space at the speed of light, which is approximately 300,000 km per second. The frequency of the travelling wave is expressed in Hertz, or Hz, which is the number of cycles or oscillations the wave goes through every second. The relationship between these three quantities is:

$$\text{speed} = \text{wavelength} \times \text{frequency}$$

Since the speed of the travelling wave is fixed, wavelength and frequency vary inversely with each other. Long wavelength waves exhibit low frequency and vice versa.

The figure above illustrates the frequency and wavelength of the entire family of electromagnetic waves. For example, the wavelength of broadcast TV and FM signals is approximately one meter, corresponding to a frequency of ~300 MHz (300 megahertz, or 300 million cycles per second). Visible light has a wavelength of slightly less than one millionth of a meter (one micrometer), giving it a corresponding frequency of 300 terahertz, or $3*10^{14}$ cycles per second. Despite their unique characteristics, one property that is common to this whole family of waves is that they all obey the Maxwell equations.

During the development of quantum mechanics in the 1920s, more fundamental understanding on electromagnetic waves was reached. The waves can be viewed as made up of clusters of almost mass-less particles, known as "photons," and the energy of each photon is proportional to the frequency of the wave. So extremely high frequency or short wavelength electromagnetic waves such as X-rays have higher energy (not to be confused with power, which refers to the intensity) and can cause damage to the body.

2

THE TELEGRAPH

Mankind has an innate yearning to discover and to innovate, and long before Faraday, Maxwell, and others elucidated the natural rules of the electromagnetic universe, many people had tried to apply electricity toward different functions. Remember the French monk Nollet? When he shocked monks and palace guards, he was in fact trying to lay the groundwork for his real goal: using electric current for long distance communications.

Many people throughout the eighteenth century had this very idea, and there had been countless attempts to devise a feasible way of putting it into practice, even before the invention of the battery. Some methods were quite ingenious. For instance, one man connected twenty-six Leyden Jars to twenty-six long electrical wires, whose ends he submerged in vats of water. When he discharged a jar, the corresponding vat of water would bubble as the wire delivered its electrical payload—thus, the man argued, notes and missives could be wired from a sender to a recipient, one English letter at a time.

Later, more sophisticated methods were devised, but technologically, many difficult practical problems remained. The birth of telecommunication using electricity was really the work of two key, competing figures: the Englishman William Cooke and an American named Samuel Finley Breese Morse. In the end, it was Morse's approach that the world universally adopted, creating the first new major industry based on electricity: telegraphy. Telegraphs represented a breakthrough in long distance communications, and the successive waves of technological innovation they brought about have rippled through to the present day.

MESSAGES SENT BY ELECTRIC CURRENT

Morse was born in the United States in 1791, the son of a preacher. Even when he was very young, he loved to paint and tinker. He attended Yale University to study fine art, and though he initially majored in landscape painting, he changed his focus to specialize in human portraiture for a very pragmatic reason: Samuel Morse was rather keen on acquiring wealth. Landscape painting didn't bring in much income in those days, but Morse found he could make a living by painting portraits for the upper crust. He typically charged $16 to $20 per commissioned portrait, which was a considerable amount of money at the time. Morse did well, but his livelihood forced him to be constantly away from home, which eventually led to a serious personal blow. During one of Morse's extended business trips, his wife passed away. However, due to the lack of robust communication tools at the time, Morse didn't even learn about his wife's untimely death until he returned, the devastating news waiting for him at his doorstep.

In 1832, the widower Morse went to the Louvre Museum in Paris. He planned to recreate many classic works in one large composite painting and then exhibit it in America, thus earning—he hoped—a small fortune. Upon his return, he was not successful in generating much income from exhibiting his painting, but he would later obtain a position as professor in the faculty of painting and sculpture at New York University, also known as NYU. While on the long ocean voyage home, however, fate intervened. Morse overheard another passenger detailing various forms of activities then being undertaken in Europe to try to use electricity in long distance communications. Recalling his own past, Morse sensed that

Samuel F. B. Morse. © Library of Congress—Digital ve/Science Faction/Corbis

there was both a need and some financial opportunity in this field, and the idea greatly aroused his interest. Though he was already forty-one years old at the time, his goal to make a fortune in art remained elusive, leaving him open to the idea of pursuing this novel communication technology.

Although Morse had never received formal training in physics or chemistry, he was very bright and highly persistent in pursuing his goal. Applying his talents as a tinkerer, Morse designed and built his first telegraphy apparatus in about two years. Unfortunately, the device was extremely crude and clumsy to use, and its communication range was limited to less than 100 feet. Morse soon upgraded his design with the help of two other people affiliated with NYU: the chemistry professor Leonard Gale and a student named Alfred Vail. Together, they created a simple but efficient telegraph machine using a manually operated switch, or a key, to enter coded messages.

The team divided their labor: Morse created the base design, and Gale helped make improvements in battery and insulation of electrical wiring. The young Vail, for his part, made perhaps the most valuable contribution of all.

Alfred Vail came from a family that owned a sizable machine-building factory in New Jersey that specialized in producing parts for steam engines. Born into the industry, so to speak, Vail excelled in mechanical design, and he was able to refine and improve many aspects of Morse's original design of the telegraph system. Still more impactful, though, was the access to his family's industry connections that Vail brought. Through his parents' investment—in exchange for 25 percent equity in Morse's nascent company—Morse was able to raise the funds to develop and produce a prototype telegraph system that could reliably transmit signals more than 1,000 feet.

Though Morse's new design represented an impressive technological leap, he had trouble gaining traction in the marketplace. People did not understand the value of the technology, nor did they feel a special need for such an application. To the general public, electricity remained the mysterious stuff of legend, like Franklin's lightning bolt. But in 1837, just after Morse had finished applying for a patent on his invention, the Treasury Department made public its intent to develop a rapid, long distance—perhaps nationwide—communications system. This objective had been influenced by the success of France's semaphore system, a message relay network very similar to the combination of beacon towers and the symbolic "flag talk" used by sailors on ships. Using the semaphore system, important news could be conveyed from Marseille to Paris within a matter of hours. The Treasury Department, sensing the value of this type of speedy communications, decided that it was high time for the United States to develop a similar capability.

When he heard this news, Morse got very excited and immediately prepared a formal proposal pointing out that his novel telegraph technology could accomplish the same ends as a semaphore system, but with far superior performance. As a proving ground, he proposed constructing a forty-mile-long demonstration system using $30,000 in funding grants from the federal government. The Treasury received a total of seventeen proposals, but only Morse proposed implementing the brand-new telegraph technology. In order to persuade Congress, Morse even made a trip to Washington to lobby for his proposal, bringing with him Gale and Vail. Together, they demonstrated telegraph operation to many government officials and a number of congressmen. They were quite impressed; however, as far as Congress was concerned, the Treasury's desire was not a matter of great importance and urgency. Despite the team's impressive efforts, without a flag-bearer inside the halls of power to champion the cause, the matter was left to die on the vine. Morse's hopes, for the moment, were dashed.

Unable to find investors in America, Morse's only recourse was to try his luck in Europe. However, although his device was met with some acclaim in France and Germany, investor interest was limited. In England, inventor William Cooke was already developing a more sophisticated telegraph system, and Morse's incursion was not welcome. Only Russia, with its vast territory and attendant need for long distance communications capability, showed any serious inclination. Still, after exhibiting his telegraph, Morse received only vague expressions of interest. He made no sales and was directed to return to the States and wait.

While back in America waiting on a response from the Czarists, Morse paid a visit to Joseph Henry, the director of the Smithsonian Institute and an internationally recognized authority on electromagnetism. Though not financially involved in Morse's endeavor, Henry shared with him a few fruits of his own research, including the concept of the "relay" or signal re-transmitter, which added to the efficacy of Morse's technology by greatly extending the distance by which electric pulse signals could be sent.

Despite this improvement, Morse at last received word after months of agonized waiting that Russia had officially declined to develop telegraph technology. Morse's efforts at finding investors in Europe had failed, and with the limited initial capital he had spent and gone, his small team was left with no choice but to dissolve. The others drifted away, and Morse shifted his own interest to the new photographic technology for portrait making. Still, his failure to succeed in telegraphy rankled him.

In 1841, four years after its original submission, Morse received word that his telegraph patent application had been approved and a patent issued. Taking this as a sign to return to the fray, Morse decided to make another effort to persuade Congress to provide a grant for developing the telegraph. In 1842, now all by himself, Morse traveled again to Washington and lobbied a new class of congressmen for funds. This time, he was successful, for two principal reasons. The first was that Morse shifted focus to rely largely on economic arguments, rather than technological ones, to persuade members of Congress. The telegraph, he pointed out, could lead to substantial rates of return for its government investors if it were successful, all while addressing a critical telecommunications need of the country. The second reason was that the simple concept of telegraphy no longer seemed as outlandish as it once had. This was not thanks to Morse, however. It was principally due to the efforts of an English inventor named William Cooke.

ANNIHILATING THE TIME-SPACE BARRIER

Europeans did not sit idly by as Morse the American pushed for the development of the telegraph. In Great Britain, William Cooke applied for a telegraph patent in the same year as Morse: 1837. Cooke's design was more elegant than Morse's, but it was also significantly more complicated, and therefore more expensive and less reliable. Despite the inherent cost and reliability problems, early telegraph adopters in the United Kingdom saw great potential in Cooke's device. In 1839, Cooke and his influential partner, Charles Wheatstone, a well-known British scientist-engineer, persuaded a railway company to install their telegraph system for use in dispatching trains along a thirteen-mile northwest London line. This marked the first commercial application of telegraph technology, and it was an instant success. A task that had previously been performed by a complicated series of signals—or worse, timetables that left no room for error, as countless train stalls and even accidents could attest—could now be accomplished via fast and reliable point-to-point communication. As it happened, there were even impressive ancillary benefits. One day, two robbers hit a London bank and escaped by jumping aboard a moving train. Not even swift horses could chase them down! However, thanks to the telegraph, the London terminus was able to send a message to other stations along the line, and when the getaway train pulled into its next stop, police stood waiting to arrest the criminals. Reports of this near-magical collar quickly spread, making the telegraph even better known. Those poor brigands could never have imagined that they would be captured through the use of this new technology!

Thanks to his arguments and the success of Cooke's telegraph in England, in 1843, Samuel Morse persuaded Congress to pass—narrowly, by a measure of

89 votes to 83—a bill to grant him $30,000 to demonstrate telegraphic communication. His proving test case was to build a cable link bridging Washington, DC, and Baltimore, a distance of forty-four miles. Upon receiving the funds, Morse immediately reunited with Gale and Vail, and he also added a new member to the team: Ezra Cornell. Gale was charged with captaining wire and battery technology as well as the procurement process; Vail was given custody of managing the telegraph transmitter-receiver equipment; Cornell was made the leader of trench digging and wire laying; and Morse himself spearheaded overall planning and coordination. However, as the project proceeded, unforeseen problems developed, leading to finger-pointing and mutual recrimination.

In truth, most of the team's problems derived from the immaturity of the basic technology and everybody's shared lack of experience. For example, the electrical wire used was made of drawn iron, and the wires' insulation was unreliable. Thus, not long after the wire was buried in the damp ground, it became rusted and unusable. Soon, nearly all of the $30,000 had been spent, but only ten of the forty-four miles of wire between Washington and Baltimore had been successfully laid.

With money and time beginning to pinch and failure a real possibility on the horizon, Morse made what became a monumental decision. To combat the cost of digging and the issue of underground corrosion, Morse proposed to hoist the electrical wires on wooden poles up in the air. As far as he was concerned, this was a purely temporary solution done in the name of expedience alone. He never once thought that, from then on, wires strung along poles would become a common sight worldwide. Cornell switched from digging trenches to setting up poles to support the wire, and the team soon finished the project. In later years, Cornell would make a fortune in telegraph system construction projects; still later, he donated some of that fortune to establish Cornell University.

On May 24, 1844, America's first telegraph line was successfully completed. Morse set up a transmitting-receiving station inside the Supreme Court building in Washington, DC, while Vail established and directed a similar hub near the Baltimore train station. Employing his well-honed public relations acumen, Morse invited the daughter of the director of the U.S. Patent Office to ceremoniously send the first telegraph, a quotation from the Bible: "What hath God wrought."

Still more noteworthy than the pomp and circumstance was the spreading of news regarding the Democratic Party's nominating convention. The delegates of the convention—which was held in Baltimore—initially nominated Senator Silas Wright of New York to be vice president, but Wright, who was in the District of Columbia, had no way of knowing this. Using the telegraph, Vail sent the message. It was delivered to Wright, who, on that very day, entrusted Morse with the task of conveying his reply: a refusal. More than anything, it was this exhibition—the rapid dissemination of news and messages that previously could have not traveled faster than a train, and the initiation and completion of a long distance exchange of information in mere hours, rather than in days—that sold the public on the genius and importance of what Morse and his men had done.

From that point forward, people squeezed around Morse's telegraph station in Washington every day, waiting to hear updates from the convention. In the beginning, some doubters only half-believed the news the telegraph was transmitting, and every item had to be confirmed several hours later by a messenger specially dispatched by train from Baltimore. After a string of successes, though, everyone came to believe in this new technology and recognize its true value. By the time the telegraph conveyed the message that the Democrats had nominated dark horse candidate James K. Polk to run for president, many were shocked and surprised by the news, but no one doubted the accuracy of the telegraphed information.

After the spectacularly successful demonstration between Washington and Baltimore, public faith and interest in the telegraph skyrocketed. Not far behind grew the fame of Samuel Morse. He became a national hero and was nicknamed "the Lightning Man," as the public still linked electricity to Franklin's experiments with lightning. After years of dutiful plodding, Morse's telegraph had reached the *tipping point.* Before that time, Morse had encountered great difficulty in convincing people of the value that the telegraph offered; after the event, however, people lined up to demonstrate their devotion to the new technology. In commemorating Morse, Congress later conferred a special honor upon him, praising him for the invention that "annihilated the barriers between time and space." Morse's invention was an unqualified success. The only question that remained was: what would be Morse's next step?

Initially, Morse hoped that he could sell his patent to the government for $110,000, from which point he imagined that the government would appoint him to lead the development of a nationwide telegraph system operated as a benevolent monopoly. However, after some deliberation, Congress decided to avoid involving itself in what it saw as the realm of private industry. Morse had failed to foresee this turn of events and was thoroughly disappointed, at least at the outset. But as private enterprise had by now become deeply interested in the telegraph and Morse held its fundamental patent, he quickly found that raising funding—plenty of funding, in fact—was no more difficult than painting a picture. After recruiting an experienced business partner named Amos Kendall, Morse decided to start raising capital for a telegraph system that would connect major cities along the Atlantic coast.

In addition to building the east coast system, which he wisely centered in New York, Morse also accepted Kendall's suggestion to license his patent to other companies who sought to build telegraphs across other parts of the nation. This, as Kendall argued, allowed the technology to become both widespread and ingrained to the point of indispensability, while not diverting Morse's company away from its own internal goals. Furthermore, there was little financial risk—indeed, Morse could simply sit back and collect royalties from the licensees! Morse saw the wisdom of this idea and implemented the licensing strategy. Soon afterwards, telegraph fever spread across the nation, with new telegraph terminals—and telegraph poles—popping up everywhere. At the end of 1846, the forty-four miles of telegraph wire between Baltimore and Washington were the only ones in the nation. Within two years, the total length of telegraph lines had increased to 2,000 miles, and by 1855, 42,000 miles of telegraph cables crisscrossed the land. In less than ten years, the telegraph network had grown to cover essentially all of the relatively densely populated areas of the entire United States, and Morse enjoyed financial gains that grew with every mile.

Many factors led to the enormous success of the telegraph in the USA; chief among these was market demand. The development of the telegraph in the United States was inseparable from the large-scale railroad construction then taking place. As in England, the use of the telegraph became essential to efficiently coordinating train dispatches between stations. Railroad companies also owned the land rights for their tracks, which eliminated potential right-of-way problems for hoisting telegraph poles. Thus, a network of telegraphy was able to quickly spread across the country alongside the railroads, creating a hub-and-spoke architecture that was later mimicked by airlines for plane routing. In the mid-1800s, the United States fast became a telegraph nation.

WIRE ACROSS THE ATLANTIC

In the ten years immediately following the founding of Morse's company, telegraph lines came to link major population centers not only in America, but also throughout Europe and some parts of Asia, Africa, and South America. But one great challenge remained: linking the world across the ocean. News could travel across land with incredible speed, but transoceanic news still plodded along at the pace of a sturdy ship fighting the currents.

Since first finding success, Morse had long been an advocate for an undersea cable that could connect the two shores of the Atlantic Ocean. And, thanks in part to the expansive development of telegraph systems and the demands they placed on ancillary industries, both battery and electrical wire technology had advanced to the point where such a vision seemed to be implementable. In laboratory testing, engineers demonstrated that telegraph signals could indeed be directly transmitted over 2,000 miles of electrical wire, close to the total distance between New York and London. Suddenly, it seemed that laying an undersea cable might be more than a pipe dream. But now, a new doubt took the old one's place: was such a project practically—and financially—feasible?

Cyrus Field, a wealthy and prominent New Yorker who had retired at the young age of thirty-three, decided that it was. After making his fortune in the paper industry, he became fixated on the idea of investing in a transatlantic telegraph cable. Despite the fact that most of the key parameters of such a project—among them, specifications for batteries, signal transmitting power, receiver sensitivity, cable-insulator properties, and even the physical method of laying undersea electrical cables—remained essentially undefined, he heedlessly went ahead and invested heavily in a transatlantic telegraph startup.

Thanks to Field's support, the cable project proceeded, with most technical know-how acquired by trial and error. As the primary financial backer, Field had substantial say in the project's management; however, the chief engineer that Field chose to oversee day-to-day operations, Dr. Wildman Whitehouse, was by training a physician, not a scientist or engineer. Being a total neophyte in telegraph technology, the man made numerous costly blunders. In spite of such shortcomings, the indomitable spirit of this team could not be denied, and after more than two years of strenuous effort and an exorbitant level of expenditure (approximately $1,400,000, at a time when the average monthly wage for workers was about $20), the first cable to link the two sides of the Atlantic Ocean was completed in 1858. Seemingly miraculously, people from London and New York could now communicate directly and instantly.

This astonishing achievement was met with great public fanfare and laudatory notes in the press. Great Britain's Queen Victoria and U.S. President James Buchanan even exchanged congratulatory messages with one another via the link. However, in the weeks that followed, the signals transmitted through the cable began to grow weaker. Line noise crept into the system, and the signals started to fade. Chief Engineer Whitehouse found himself stumped and, perhaps believing the issue was one of inadequate voltage, he suggested increasing the battery potential to an improbable 2,000 volts. This desperation move did not improve system performance; instead, the line fell totally silent and no more signals came through. The transatlantic cable more than two years in the making lasted less than three months and transmitted a scant 732 messages before atrophying and dying completely.

As elated as they'd been by the initial success, the public was greatly discouraged by this failure. American and British governments, too, were disheartened by the affair. A special team consisting of eight experts, four from each country, was formed to determine the cause of the failure. The British team was led by none other than James Clerk Maxwell's old Cambridge mentor, Professor William Thomson. He shepherded the joint team through a deep and thorough analysis that clearly and succinctly pointed out the various design flaws that existed within the original system. Professor Thomson also developed a new signal receiving technique that greatly enhanced receiver sensitivity.

Lord Kelvin (Prof. William Thomson). Science Source

After listening closely to Professor Thomson's analysis and suggestions, Cyrus Field decided to invest an additional $2,500,000 in constructing a second telegraph cable across the Atlantic. This time, he wisely vested Thomson himself with full control over the cable system's technical design and implementation. Thomson and his team, learning from the mistakes of the first effort, methodically and comprehensively applied their scientific knowledge in designing every detail of the system. Thanks to their work, the second transatlantic cable, completed in 1865, was an unmitigated success.

From that time on, communications flowed smoothly and reliably between the two sides of the Atlantic. Those involved with the project were amply rewarded: Queen Victoria ennobled Thomson as Lord Kelvin, a great honor for the world of science in Britain. For his part, Cyrus Field earned back his investment within three years. The transatlantic telegraph service initially charged $1 per letter for sending and receiving messages; the operation was so profitable that the net earnings on the first day alone totaled over US$4,000, a sizeable sum in that time.

Concurrent with the success of the transatlantic cable, more of the globe became wired by the telegraph. In Europe, Werner Siemens constructed Germany's domestic telegraph system, and his company later helped Russia construct a link from the Baltic Sea in the Nordic lands all the way to the Black Sea on the border of Ukraine. Finally, in the 1870s, Siemens even managed to complete a telegraph link from London to Calcutta, slashing the delay in Britain's communications with the jewel of its empire from one month's time to mere minutes. With telegraph systems connecting a large portion of the globe, mankind for the first time had harnessed the special functions of electricity and magnetism to provide fast and reliable long distance communications. And this appeared to be only the first hint of what electromagnetic technology could do.

INTELLECTUAL PROPERTY DISPUTES

One unexpected problem that arose during this widespread sprouting of telegraph technology was the issue of patent rights. Patent rights, also known as intellectual property rights, are the statutory rights of inventors that allow them to safeguard their creations. Patented inventions are legally protected over a fixed period of time, during which no other party can use (or "infringe upon") them without the rights owners' consent. Often, inventors choose to permit other people to use their inventions in exchange for a royalty fee for the right of usage. Morse, in implementing Kendall's licensing scheme, was among these. This marked the first time that large-scale licensing of a new technology was practiced, and it was a good test case for the effectiveness of patent law enforcement.

Patents have been extremely important in the commercialization process of new technologies. The system was first created in 1474 in Venice during the early stages of the Renaissance to provide economic incentive to inspire inventors. The first patent for a technological innovation was granted in 1499 to John of Utynam for his method of manufacturing stained glass for church windows. In America, the first patent law was created in 1790 under the direction of Thomas Jefferson, and its essence was written into the Constitution.

Morse, on his voyage from Europe back to America in 1832, heard about the possibility of using electric current in communications. He later developed his own telegraph technology and in 1837 submitted a well-constructed patent application. That patent was granted in 1841. When the U.S. government decided to let private enterprise develop telegraphy, Morse followed Kendall's suggestion and licensed his patent rights to several companies for an attractive royalty fee. And beginning in 1845, telegraph development sprang up everywhere as companies both big and small invested in this business.

Many companies legitimately followed patent law in developing America's telegraph network. They came to an agreement with Morse and paid him for the right to use his technology. However, many more simply ignored Morse's patent and built their own systems anyway. Even some among those that contracted to license the patent legally came up with various excuses for not paying their royalty fees. Under such circumstances, Morse found himself with little choice but to sue these companies for infringement. They, in turn, responded by banding together and challenging the legality of Morse's patent in court. Thus, in the late 1850s, the hero Morse—the Lightning Man—found himself embroiled in a series of bitter trials. His opponents charged that much of Morse's technology was lifted from other, earlier advances. They even persuaded Joseph Henry, America's foremost authority in electromagnetics and the man who had taught Morse about relay transmission, to testify that Morse had purloined his invention.

The case eventually reached the U.S. Supreme Court, which, in 1854, decided that Morse was the inventor of the telegraph, ruling in his favor in the civil litigation. The decision stated that although Morse had adopted some knowledge of his predecessors, this did not diminish the fact he was the first to successfully integrate such knowledge and apply it to communications, thus making him the legal inventor of the telegraph. This ruling established Morse's place in history and brought him considerable wealth, and it also set the bar for future intellectual property cases. Even today, struggles over patent rights for many new inventions are commonplace and more intense than ever before.

Historically, innovation has often involved the contributions of many people, and it can be difficult to determine just who the primary contributor is. It is commonly accepted, for instance, that Columbus discovered the New World of America. In reality, Columbus actually never went beyond the Caribbean islands, and the Viking chief Leif Ericsson set foot onto the North American mainland some 400 years before Columbus. However, Ericsson's stumbling on and off the coast of Newfoundland had no meaningful consequence to the history of mankind, and in considering the impact of the two men's discoveries, Columbus' was immeasurably the greater one. Therefore, there is little dispute over the notion that credit for the Europeans' discovery of the Americas should rightfully go to Columbus. Similarly, Morse was the central figure in the invention and promotion of the telegraph, and no one could claim any greater contribution. Therefore, the dispute over telegraph patent rights was readily settled in his favor.

In any major technical innovation, the typical process first involved is identifying and defining a user need. A solution is then formulated that involves the creative integration of multiple technologies, including the possible use of other people's existing inventions. Finally, a proposed solution must be successfully implemented in a form that is economically viable and creates the most value for the end user. In the telegraph case, Morse clearly had accomplished all these requirements, and he therefore deserved the recognition as the leading innovator. However, Morse deliberately failed to acknowledge and reward key contributions made by other people, including Vail and Gale. Morse was snobbish and covetous, never willing to share credit for his accomplishments. In the end, almost all the individuals who ever collaborated with him rancorously broke off ties, and many sued him as well.

MORSE CODE

As the vanguard for widespread application of electrical technology, the telegraph was and is necessarily quite simple. The system hardware consists of the battery, the telegraph key or switch, the electrical wire, and the electromagnet. When the telegraph key is depressed (or the switch is closed), this completes a circuit, allowing current from the battery to pass through the long wire and flow to the electromagnet. The electromagnet is used to detect the presence or absence of electrical current; the signal can only transmit the "on" or "off" state of the key.

These discrete binary signals were known as "digital signals," from the Latin root for two. The challenge became the question of how to transmit message content using only these two signal levels—in other words, how could the telegraph "digitally code" the message? During the course of developing their telegraph systems, both Morse and William Cooke came up with their own coding schemes, but in the end, Morse's code was universally adopted because of its simplicity.

Interestingly, Morse used and abandoned two different preliminary coding methods during the course of telegraph development before finally hitting upon his third, best, and final methodology. In the first approach, he developed many fixed, unique combinations of "on" and "off" states to represent some commonly used English words. However, this method proved too limited, and it was quickly discarded. In his second iteration with Vail on board, he developed a coding scheme that used ten different combinations of "on" and "off" signals to represent the digits 0 through 9. He then grouped quartets of digits, allowing him to transmit numbers from 0000 to 9999, each of which referred to one of 10,000 commonly used English words he codified in a special book for telegraph operators. This was the method he used when he first lobbied government officials in Washington, DC, in 1837. However, this method required time-consuming and unwieldy hunting in the code book for translations of ordered numerals, and it, too, was abandoned within just a few years of its inception.

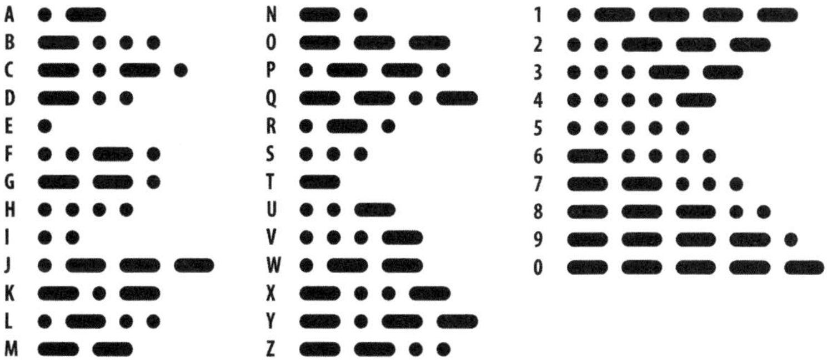

International Morse Code. Derek Cheung

The coding method finally adopted for the telegraph was a series of digital pulses that represented thirty-six distinct combinations, each uniquely corresponding to the numerals 0 through 9 and the letters A to Z. This system is commonly known as "Morse code." In Morse telegraphy, the single telegraph key has only two positions, on or off. When on, the key can be depressed for either a short burst (a dot, [·]) or a long one (a dash, [-]). By grouping dots and dashes into standardized combinations, Morse's telegraph could readily transmit and receive messages, letter by letter (rather than word by word), that were easily coded and decoded by telegraph operators with only modest training and without using a code book.

Morse code was successfully demonstrated in the 1844 Baltimore-to-Washington experiment. Interestingly, Morse had not spent much of his own time improving telegraph technology after his initial patent application in 1837. Thus, it is very likely that Vail, and not Morse, played a major role in developing the final version of "Morse's" code. In fact, when Vail died in 1859, his widow immediately sued Morse for stealing her husband's invention. Morse, however, never gave Vail any credit, and he even reneged on the 25 percent equity share of the business he had promised to Vail's family upon their initial investment. The legal case was still active when Morse died in 1872 at the age of eighty-one. Perhaps the famed Morse code should really be known as Vail code?

IMPACT

One major advantage of digital technology was the simplicity of its transmission. As electrical signals propagate along a wire, their signal strengths attenuate and their shapes become distorted and stretched out. In a digital system, such problems are easily remedied by using a relay, which generates a new signal as soon as original signal strength attenuates below a certain threshold. This technology, first pioneered by Henry, allowed telegraph signals to be transmitted and regenerated over long distances without losing their accuracy.

Similarly, the quality of electrical wiring improved exponentially as time and demand progressed. The earliest wires were made of iron, which would easily rust; besides, they were poorly insulated and their diameters were uneven. Later, these haphazard strands evolved into highly-insulated, low-loss, isometric copper wires. The development of the transatlantic telegraph brought electrical wiring technology to a still-higher level of sophistication and maturity. During the development of the second transatlantic telegraph cable, scientists and engineers working under Lord Kelvin, then still known as Professor Thomson, analyzed in great detail every aspect of cable technology. For example, as a member of the technical team, German engineer Gustav Kirchhoff developed a new theory based on calculus to analyze and predict the attenuation and distortion of telegraph signals as they were transmitted along the cable. His studies later evolved into "circuit theory," the foundation of modern electronic circuit design.

The rapid growth of the telegraph industry also spurred the inexorable development of many new products and manufacturing technologies. Batteries, for example, had been used primarily only in laboratories for experimentation, but with the development of telegraphy, they became a huge and important commercial market. And as the demand for batteries increased, battery technology advanced by leaps and bounds, evolving from the primitive Voltaic Pile into the powerful Daniel cell, and later, to the advanced dry cell and the lead-acid rechargeable. With each improvement, the battery became more compact and boasted ever-higher stored energy density. Still, these batteries had their limitations, and by the late 1870s, telegraph companies would begin to use newly invented dynamos—electrical generators—to power their systems. Finally, as telegraphy continued to develop at a breakneck pace, a number of specialized companies were formed. Some companies operated as service providers, sending and delivering telegraphs for their customers, while others manufactured telegraph equipment or designed and constructed new telegraph systems. From the late 1840s through the end of the century, the whole telegraph industry was booming.

In the early days of telegraph service, many companies of all sizes played in this space, serving one or multiple functions. In due time, however, the larger companies quickly seized competitive advantage. Once any individual company had control of the bulk of a section of the network, it became extremely difficult for other companies to compete.

By the end of the 1860s, a megalithic company named Western Union had staked its claim to much of the telegraph network. The company was originally founded by Ezra Cornell, the one who helped Morse to build the first telegraph system. Later, the company was able to hire William Orton, an outstanding businessman with great vision and management skill, to be its president. Orton was among the first to employ the franchising model, as used widely in today's fast food industry and other business chains, to rapidly expand his telegraph business. By 1873, Western Union controlled 90 percent of America's telegraph traffic volume, and it had at the same time developed a vertically integrated business structure that integrated telegraph services, equipment production, and new systems construction under one roof. Using its own telegraph infrastructure as its communications backbone, Western Union pioneered the business practice of closely coordinating the operation of multiple remote physical locations. Its reach was so broad and its market position so dominant that in 1884, Western Union became one of founding eleven companies to form the Dow Jones Transportation Average Index.

Because of its growing ubiquity in the international landscape, the telegraph became the focus of an expanding range of telecommunication applications. The military seized the value of telegraph early, and first used it during the Civil War. Telegraph played even a larger role during the Crimean War in 1853–1856, when armchair generals in London and Moscow could send immediate orders to field commanders at the front line, which forever changed the nature of war. But by far the largest users of the telegraph were financial, commercial, and news-oriented institutions; the telegraph began to find itself at the center of an ever-expanding information industry. For example, prices on the New York commodities and stock market exchanges had always fluctuated quickly, sometimes changing significantly in a flash. Before the telegraph, many trading companies employed swiftly running boys whose sole purpose was to rush at high speed to and from the trading floor with the most up-to-the-minute prices. After the installation of the telegraph, however, trading companies no longer needed the services of these hustling urchins—instead, they relied upon a dedicated telegraph line to their premises that could provide pricing moment by moment. Sensing a new opportunity, a young inventor by the name of Thomas Edison devised a method by which telegraphs could continually and automatically update and report on prices of tradable goods on the markets: the ticker tape. This invention quickly proved to be a success, as any commercial trading company not employing it quickly found itself at a serious competitive disadvantage. Thanks to his ingenuity, Edison made his first fortune.

Next emerged the service of using telegraphs to "wire" money. In 1871, Western Union, the national giant, became the first to offer such services, using communication technology to facilitate more rapid and effective business

Edison's Ticker Tape. Courtesy of Spark Museum

transactions between financial institutions. Today, the telegraph is in disuse and Western Union is far from the titan it once was; however, the service of wiring money remains in place throughout the world, especially in developing countries.

Of course, the telegraph's use in commercial communications was not limited to the financial industry. News—the heart of mass communication itself—once relied on physical transport by horses, boats, or trains. Directly delivering documents to far-off lands could easily take as long as several months, leading to situations in which "fresh" international news arrived already stale. Thus, the press and news agencies became early and eager clients when long distance telegraphy services became available.

There was, of course, a downside to relying on the telegraph to transmit news: cost. Telegraph fees were fairly expensive, with rates set at a per-word level. Thus, the cost of transmitting a long piece of news was very high. Inevitably, in 1848, several New York news companies formed an alliance to consolidate the news they had collected from various places into one pooled source. This source they formed became known as the Associated Press (AP). Using its collective purchasing power, the AP was able to negotiate the lowest possible price with the telegraph companies, allowing the news syndicate to disseminate information widely while keeping economic overhead cost manageable. Similarly, the European financial news company Reuters, which had long relied heavily on the homing pigeon, established a mantra of "Follow the Cable." Reuters abandoned the birds and moved its head office to London, then the world's foremost telecommunications hub.

The period spanning from 1844, when Morse made his first demonstration of telegraph communications from Washington to Baltimore, to 1865, when the second transatlantic cable charged into service, lasted a mere twenty-one years. In that brief span, the telegraph managed to cover the whole world, changing people's lives for good. Due to the popularization of telegraphy, the public's latent dread of electricity and magnetism gradually faded and was replaced by familiarity and admiration. In 1865, the International Telegraph Union (ITU), history's first global technology organization, was formed to regulate and standardize the world's telegraph systems and technology. Not surprisingly, the ITU decided to standardize global telegraph operations based primarily on the Morse system.

3

THE TELEPHONE

The story of Alexander Graham Bell's invention of the telephone is legendary. On March 10, 1876, when Bell was carrying out an experiment in his laboratory to try to convert voice signals into electrical current, he inadvertently spilled acid onto his clothes. He shouted for his assistant in another room, "Mr. Watson, please come here! I need you!" Watson heard Bell's voice, not through the air, but from a signal transmitted through the electrical wire. With this, Bell found that he had successfully invented the telephone.

It's a nice tale, and parts of it hold water. However, there are a number of disparities between this story and the real, factual, historical circumstances behind it.

VOICES CARRIED OVER WIRE

By the 1870s, the telegraph had become an indispensable means of communication. By the hour, it seemed, new telegraph lines were being constructed, and there appeared to be no end in sight. But as construction expenses grew and line costs rose commensurately, engineers began to wonder whether two different telegraph signals, or telegrams, could be sent at once along the same line. If someone could indeed devise a method for sending multiple signals simultaneously, their work would serve as a quantum leap for the nascent telegraphy industry. Soon, someone did invent a way to send two telegraphs on the same line; the young inventor, Thomas Edison, even came up with a way to send four messages at once. However, these approaches were quite limited and could not be scaled beyond what had been accomplished already.

One elegant and effective way to transmit a large number of telegrams through the same line would be to transmit using different tones for different messages. For example, if one telegraph message was transmitted with a low pitched "Da, Da, Da," and another was transmitted at the same time with a high pitched "Di, Di, Di," these two messages would not commingle. Either a man or his equipment could readily distinguish between these tones. This principle, known as "frequency multiplexing," could accommodate a large number of separate messages sent simultaneously. In fact, the principle is used extensively even today in communications technologies: radio station settings and TV channel transmissions are differentiated by nothing more than a slight change in the tones that carry the information content.

The basic principle of frequency multiplexing is easy to understand and appreciate. But how were inventors able to implement the idea in practice? This was a significant technical challenge in the 1870s. At the time, there were two individuals most interested in attacking the problem. The first was a veteran inventor named Elisha Gray, who held over seventy patents and was the founder of Western Electric Company, a company that specialized in manufacturing telegraph equipment and which was later acquired by Western Union. The second was a very bright young man with far fewer credentials; his name was Alexander Graham Bell.

Elisha Gray. Courtesy of Highland Park Public Library

Bell was born in Scotland in 1847. His father and grandfather were both celebrated specialists in speech research, and his mother was deaf. From childhood, Bell was strictly trained in pronunciation and elocution, and he became accomplished in declaiming

Shakespeare verses and playing classical music, especially on the piano. He was a refined young man of cultivation and charm, and because of his mother, he was filled with a sense of compassion for deaf-mutes. As he grew older, this blossomed into an academic interest in the principles of speech and the basic physics of acoustics.

When he was seventeen years old, Bell immigrated along with his parents to the New World, settling in southeastern Canada. Several years later, he left his home and moved to Boston, Massachusetts, where he became a private teacher for deaf-mute students. Among his pupils was a teenage girl named Mabel Hubbard. Although they were eleven years apart in age, the two became quite fond of each other. Mabel's father, Gardiner Greene Hubbard, was a well-known attorney and an expert in patent law. Mabel's mother was from the most prominent and wealthy family in Boston. Hubbard, with his close personal connections in both Boston's upper society and Washington's legal and government circles, was a man of considerable influence. He had expressed the view that Western Union's domination over the telegraph industry was bad for fair market competition and should be broken up. He even proposed to nationalize the company and operate it as a semi-private institution under the auspices of the U.S. Postal Service, with him helming day-to-day operations. Though Congress failed to support this somewhat self-serving measure, Hubbard remained deeply interested in the telegraph industry and fascinated by any new technology that might impact it.

Alexander Bell in 1876. Courtesy of AT&T Archives and History Center

Bell's interest in acoustics and Hubbard's bent toward the telegraph yielded common ground, and at a dinner at the Hubbard home in the summer of 1874, Bell, hoping to leave a favorable impression on Mabel's family, expressed some of his ideas to the family patriarch. Drawing on his knowledge of music, he explained how, by using the method of resonance, one could generate different harmonic tones from the same vibrating string of a musical instrument. This principle, he argued, could be applied to send multiple telegraphs concurrently through a single wire! Hubbard's interest was aroused by this, and he decided, along with another wealthy Boston businessman, to financially support Bell's development of this "harmonic telegraph" technology.

Utilizing the investment from Hubbard, Bell leased laboratory space near downtown Boston and hired a technical assistant named Thomas Watson. Bell was clever and full of ideas, though not very skillful or well versed in carrying out hands-on experiments. Watson, on the other hand, was a trained engineer with experience in both electrical and mechanical design. The two men's strengths complemented each other, and they worked together very well. However, their technical progress was slow, and after over a year of effort, they had little to show for their work.

Gardiner Green Hubbard. Courtesy of AT&T Archives and History Center

Time dragged on, and Bell remained unable to achieve any technological breakthroughs. However, he did find himself reaching new heights in one area: affairs of the heart. As the seasons progressed, Mabel Hubbard and Bell found that their feelings for one another deepened until, at last, they had fallen hopelessly in love. Near the end of 1875, Bell gathered up his courage and revealed to Mabel's parents his deep attachment to their daughter. He expressed, too, his wish to spend his life with Mabel and asked for her hand in marriage. Gardiner Hubbard was unmoved by this display, however, and he remained noncommittal. His ambivalence, it became clear, hinged upon Bell's inability to make significant progress with the harmonic telegraph. He was reluctant to place his daughter's life in the hands of young Bell, who up to that point had not accomplished anything significant.

Bell found himself under tremendous pressure. He tried everything he could think of, even seeking counsel from America's foremost authority on electromagnetics, old Joseph Henry, for help with his technological problems. Now well-advanced in years, Henry couldn't offer much specific help; he simply encouraged Bell to redouble his efforts. During this time, Bell did make a conceptual breakthrough: with his knowledge in music and acoustics, he realized that if he could transmit different electrical tones through a wire, then it would be possible to transmit voice and music as well, since they could be viewed as composed of many different tones with varying amplitude and timing. When he expressed this revelation to Hubbard, he was told to stop being distracted by these peripheral thoughts and to instead focus his energy on solving the harmonic telegraph problem at hand.

Hubbard was a pragmatic man, and more than anything, he wanted Bell's harmonic telegraph idea to succeed—not least because of all the investment he'd already made. Through his private connections among Washington functionaries, Hubbard learned that Elisha Gray of Western Electric was himself very close to successfully developing a harmonic telegraph and would soon be ready to apply for a patent. Hubbard turned up the pressure on Bell and told him to write up all his thoughts and findings, even if incomplete, into a patent application that could be submitted at a moment's notice. Bell, though still facing technological hurdles, did exactly as he was instructed.

On February 14, 1876, Hubbard learned from a source inside the Patent Office that Gray had submitted a provisional patent with the title "An Apparatus Using Telegraphic Means to Transmit and Receive Sounds." U.S. patent law at the time stipulated that inventors must possess successful experimental results before a patent could be awarded; a provisional patent would allow an inventor to apply for a patent based on an idea not yet reduced to practice, provided he could produce proving experimental results within twelve months of filing. Upon hearing that Gray's provisional patent had been received, Hubbard immediately sent Bell's pre-prepared patent application to the Patent Office. He then illegally paid off an employee there to have Bell's application recorded as having been received a few hours earlier than Gray's. The title of Bell's patent application was "Several Improvements on the Telegraph," which was fairly general and vague. It included only a few nondifferentiating thoughts on the harmonic telegraph concept, and it included no experimental results at all.

On February 24, 1876, Hubbard arranged for Bell to go to Washington, DC, and secretly and illegally meet with a patent examiner, who let Bell look over Gray's application documents to learn Gray's idea of converting sound signals to electrical current. He also allowed Bell, in the margins of his already-submitted application, to add a new and unrelated claim about inventing a "Means for a Variable Electrical Resister to Produce Undulating Electric Currents," the text and theory of which were copied straight from Gray's application. This idea would prove to be the core technology behind the successful invention of the telephone, yet it was completely unrelated to other parts of Bell's original application document.

On March 7, 1876, shortly after Bell returned from Washington, he got a notice from the Patent Office that his patent application had been approved. This was extremely unusual, since the patent process normally took many months, or even years, as with Morse's telegraph patent. Furthermore, Bell had not yet demonstrated his claims, and no official hearings had been held to resolve the overlaps with Gray's provisional patent application. Of course, the clandestine and unlawful meetings—to say nothing of the possible bribery involved—went unannounced.

With the granted patent in his pocket, Bell immediately went back to his laboratory and turned all his attention to trying out Gray's idea of using a liquid-acid based variable electric resister. Sketching out an experiment in his log book identical to the one detailed in Gray's application, on March 10, 1876, Bell succeeded for the first time in transmitting sound signals through wire using electric currents. When he carelessly spilled some acid from a beaker onto his clothes, he immediately called out for Watson's help, expecting not a telephonic miracle but only his shouts to carry his voice.

The news of Bell's success in transmitting voice signals through a wire spread, but how it was done was kept a strict secret. To Bell's surprise, he even got a congratulatory note from Elisha Gray.

Hubbard, the ever-enterprising businessman, pushed Bell to demonstrate his technology to the public. He especially wanted Bell to show his device at the 1876 International Centennial Exposition to be held in June in Philadelphia, an enormous World's Fair held to commemorate America's one-hundredth anniversary. But Bell was extremely reluctant. How embarrassing it would be to publicly flout the supposed development of obviously pirated technology! But Bell had no choice; as Hubbard exerted his pressure through the promise of Mabel, Bell knew he had to do something—and he had only three months to make it work.

Perhaps cognizant of the fact that it might look odd to suddenly keep detailed records when there was no prior recorded evidence of his "work," Bell kept very limited notes over the ensuing three months. Nonetheless, based on Gray's concept of using a variable resistor to convert acoustic waves to electrical current, Bell was able to implement the basic idea using a different, magnet-based method. Though the voice quality of the magnetic transducer was no better than that of the liquid-acid design he had stolen from Gray, it was nevertheless original and adequate to get Bell through public demonstrations. Elisha Gray even witnessed Bell's demonstrations at the Fair and did not suspect that Bell had purloined his invention for the original demonstration.

Bell was so relieved after the demonstration that he wrote Mabel "I wish the whole telephone business were off my hands altogether . . ." Shortly thereafter, he learned that the German-American inventor Emile Berliner had invented a new sound wave-to-electrical current transducer with much better performance than his design, and Hubbard immediately bought the technology. Armed now with a basic patent for telephone and a capable transducer technology, Hubbard was ready to launch a new business. As for Bell, he was only too happy to walk away from the telephone to pursue his true interest: his love. With her father's blessing, he wedded Mabel Hubbard at last, and the newlyweds took an extended honeymoon to Europe that lasted over a year and a half while Hubbard pondered how to monetize the value of the telephone patent.

BUILDING THE TELEPHONE BUSINESS

Though Hubbard was an old hand at political maneuverings and legal battles, building up a business from scratch was not his strong suit. After some initial considerations, he decided, to everyone's surprise, to approach the entity against which he'd once so agitatedly railed: Western Union.

Hubbard approached William Orton, the president of Western Union, and offered to sell him Bell's telephone patent for $100,000. Orton assigned the company's most senior technical staff to review the telephone technology, but their conclusion and advice was that Western Union should pass. With its extremely weak and poor sound quality, they claimed, the telephone was little more than a toy! Furthermore, they posited, Western Union already controlled the most advanced telegraph technology in existence, and there was no need for them to develop a talking telephone. This kind of reaction to outside technologies was and is quite common amongst big, established companies, where commitment to legacy technologies and internal conflicts of interest can impede objective perspective and decision making.

After Orton read the report, he turned down Hubbard's offer without further reflection. Lacking any better ideas, Hubbard began to consider his other options. He eventually decided that his best course of action would be to start a company to develop the telephone business himself. Naturally, he wanted to attract top-quality start-up and management talent, and for that, he had just the candidate in mind.

Theodore (Ted) Vail was the son of a distant cousin of Alfred Vail, the man who, a few decades earlier, had helped develop Morse's telegraph. As a young man, Ted Vail was entrepreneurial and a strong leader. He left home at an early age to earn money as a telegraph operator. Later, he applied his telegraph-based network expertise in developing logistics and route scheduling techniques for efficient mail delivery by the railroad companies. After he made a name for himself, he was promoted and assigned to Washington, DC, to take charge of supervising the national rail-postal service. In this capacity, he was called upon to interface with high government officials, congressmen, and lobbyists, as well as top business leaders in banking and transportation. As he demonstrated, he had excellent interpersonal skills as well as the management smarts needed to excel in this post. Hubbard was very familiar with the goings-on in Washington, DC, and, in all his contacts with Ted Vail, he had emerged deeply impressed with this outstanding young man.

Theodore "Ted" Vail. Courtesy of AT&T Archives and History Center

In early 1878, Hubbard talked Ted Vail into leaving Washington to join the newly founded Bell Telephone Company and help grow the business. Ted Vail was given the title of general manager, despite the fact that he was just thirty-two years old. Nonetheless, with his unique management abilities, financial skills, and his ability to leverage Bell's telephone patent, Ted Vail was able to steadily build Bell

Telephone Company into a successful enterprise. Still, unbeknownst to Vail and Hubbard, dark clouds were gathering force, and a powerful competitor was emerging on the horizon.

PATENT BATTLE OF THE CENTURY

It was not long after William Orton turned down Hubbard's offer of the telephone patent that he realized the seriousness of the mistake he had made. As the business of Bell Telephone grew, Orton began to see the tremendous market potential of the new technology. He determined that Western Union had to get into the emerging telephone business right away. Rather than return to Hubbard hat in hand, however, and make a new, larger offer for the telephone patent, Orton decided to challenge Bell Telephone directly in the marketplace.

Despite his position as a late entrant to the telephone market, Orton possessed what he believed to be three distinct opportunities for advantage. The first was the sheer size of Western Union's existing telegraph network infrastructure, which could be modified to carry telephone messages. That alone intimidated most other players in the market. The second was Orton's belief that due to Elisha Gray's independent and original work on voice-over-wire technology, Bell's patent might not be able to block Western Union's entry into the market. The third and final toehold apparent to Orton was related to technology: even though Bell Telephone was using the improved current-to-voice transducer purchased from Berliner, the absolute performance level of the Bell Company's telephone was still poor. For example, it often dropped consonants in words when audio signals were weak. Orton believed that if Western Union could develop a better transducer, he would be able to create a plainly superior telephone and therefore offer better voice service. But who should Orton turn to for coming up with such an invention? As it happened, Orton had an idea, commissioning the task to a budding genius who had already found repeated success in new telegraphy technologies: Thomas Alva Edison.

Orton gave Edison $100,000 for the development contract, the very amount he'd refused to pay Hubbard earlier on, along with instructions to create an improved voice-to-current transducer technology. In three months, Edison managed to create a high-quality transducer that he called a "microphone," or a miniaturized telephone, which performed far better than either Gray's pilfered liquid-based variable resister or Berliner's electromagnetic transducer.

Between the clarity of Edison's microphone and the leveraged strength of its existing wired network, Western Union's telephone business—organized as a subsidiary company in which Elisha Gray, telephonic researcher and former owner of Western Electric, owned 33 percent—rapidly took off. Soon, it was closing in on Bell Telephone and constituting a major threat to that company's survival. A panic-stricken Mabel Hubbard Bell and her husband, Alexander, quickly sold off a number of their shares. Ted Vail and Gardiner Greene Hubbard realized that they, too, faced ruin, and after discussing their options, they decided to take the only option the company had to survive: in 1878, the still-tiny Bell Telephone Company sued behemoth Western Union for patent infringement in open court. This suit would become known as the famous Dowd case of 1878–1879.

At the time, the case caused something of a media conflagration, a firestorm that engulfed the interest of the entire country. Western Union was resolute that they did nothing wrong. They maintained that Alexander Graham Bell had not been alone in working on harmonic telegraphy, as Elisha Gray had done much work independently. They also made clear that their actual telephone was a grand technological improvement over Bell's. Across the aisle and under instructions from Hubbard and Vail, Bell Telephone's lawyers stuck to a few key principles: Bell's patent was an officially issued patent, while Gray's application was provisional, and Bell's application was received at the Patent Office earlier than Gray's. They managed to discredit the original patent examiner as a witness and successfully blocked the request by the defending lawyers to submit the original copy of Bell's lab notebook. It was quite shocking that the pillars of telephonic history were being constructed on such an underhanded and shaky foundation.

Throughout the events that led to the first transmission of voice through the wire, Hubbard's was the main hidden hand. He orchestrated and maneuvered time and again to ensure he came out on top, and he did not relent in his activities during the trial. Exerting pressure on his son-in-law, he even induced Bell to betray his own conscience and testify in court that he had designed the very mechanisms he knew that he had stolen. Though Bell's gentlemanly demeanor, verbal eloquence, and sincere image left a good impression on the court and the public, compounding this lie caused Bell to suffer tremendous psychological stress. For love (or at least the approval of his beloved's papa), Bell was willing to comprise his integrity, commit thievery, and cover up his father-in-law's lies. Of course, he never did consider the damage his craven actions caused to Elisha Gray.

Hard as it is to admit, history often hinges on chance, and the case of the telephone is no exception. At the height of trial, with the jury deadlocked, something completely unforeseeable happened: William Orton died.

Orton had dominated Western Union for over fifteen years. He had been the shepherd who successfully oversaw its growth to dominance and transformed its operation, and his death had a massive impact on Western Union's strategic business direction. Additionally, as a publicly traded company suddenly suffering from a leadership vacuum, Western Union found itself in the crosshairs of a Wall Street financial concern backed by railroad baron Jay Gould that specialized in hostile takeovers.

The remaining management of Western Union saw the future of the telephone business in quite a different light from how Orton had, and what's more, they needed to divert their full attention from the court battle—in which they were named as defendants—to driving off the corporate raiders. So in June 1879, with the trial still in full swing and the outcome uncertain, Western Union approached Bell Telephone with an offer to settle their case out of court.

The accord reached was as follows: Western Union agreed to withdraw from the telephone market and sell to Bell Telephone all its telephone assets, including its installed subscriber base of 56,000 as well as Edison's microphone patent. At the insistence of Bell Telephone, Western Union also formally conceded that Alexander G. Bell was the inventor of the telephone. In return, Bell Telephone agreed to assign to Western Union 20 percent of all income from its telephone patent over the next seventeen years, 40 percent of its equity in local Bell Telephone companies in New York City and Chicago, and it agreed not to enter the telegraph market.

At outset, both sides appeared to have made some concessions: Vail got free reign in the telephone market for Bell Telephone, while Western Union emerged in control of telegraphy with management's dignity unscathed, gained significant equities in several important local Bell companies, and seventeen years of free telephone royalties to boot. Of course, time showed that Bell Telephone came out ahead in the deal. Consolidating the telephone business of its main competitor, Bell Telephone began a rapid growth phase. Within two years, Bell Telephone even acquired Western Electric from Western Union, employing the company's manufacturing strength to exclusively produce telephone equipment for Bell Telephone.

It was not until years later that the truth of Alexander Graham Bell's theft of Gray's invention came to light, and when it did, Gray found himself furious, resentful, and wholly without recourse. Western Union had already legally acknowledged Bell as the inventor of the telephone, and Western Electric, the company Gray himself had created, had fallen into Bell Telephone's hands. All that Gray could do was sit back and die a bitter man in 1901. Shortly thereafter, a note was found among his personal effects that is perhaps the wryest of all commentary on the birth of the telephone. It read, in part, "The history of the telephone will never be fully written. . . . It is partly hidden away, and partly lying on the hearts and consciences of a few whose lips are sealed—some in death and others by a golden clasp whose grip is even tighter." The original note was in Gray's archive at Oberlin College.

Both Western Union and telegraphy eventually began to decline in importance. Telephone technology, however, swelled, and over time, Vail grew and transformed Bell Telephone into American Telephone and Telegraph, or AT&T, which dominated the U.S. communications businesses for over one hundred years until its federally mandated breakup in 1983. Vail also established Bell Laboratories, which was the cradle of modern electronic and information technology and allowed AT&T to remain the world leader in communications technology for generations. This path was aided, in part, by Vail's adept skills as a political strategist. As AT&T grew, Vail persuaded the government that in order to ensure a universal, nationwide, high-quality telephone service, AT&T must be allowed to operate as a "benign monopoly." He willingly accepted government supervision and provided full cooperation in order to ensure that no company ever really challenged AT&T's primacy—and the government agreed. Peter Drucker, the American guru on modern business management, said, "Ted Vail was one of the greatest managers in American corporate history." And yet, very few people today have heard of him or know of his contributions. Of course, history, as they say, is written by the victors, and Vail and his public relations machine were quite successful in mythologizing and cementing the legacy of one Alexander Graham Bell as the icon of the company. All the while, Vail carefully played down his own role.

In addition to Bell, Vail, and Gray, many other people influenced the early technological development of the telephone, including the Englishman Francis Ronalds (1816), the French soldier Charles Bourseul (1854), the German Johann Philipp Reis (1861), and the Italian-American immigrant Antonio Meucci (1862). But their contributions only indirectly impacted the subsequent development and popularization of the telephone. We can refer back to the case of Leif Ericsson and Columbus and credit the creation of telephone technology to Gray and Bell, and the business success of the telephone to Ted Vail. To bring an invention to prominence in the market and build a business around it, many factors are needed beyond technology, including financial investment, skillful management, and timing. Management talent like Vail's is often as important as the invention itself in ensuring business success, and such talent is most sought after by venture capital investors to lead today's new technology-inspired start-up companies.

SOUND OF MUSIC

There were many different technologies involved in the development of the complete telephone system, including network design and automatic call switching.

However, the core enabling innovation was the conversion of sound waves into electrical signals and vice versa. How was such a feat performed?

Gray's initial invention called for fixing a conductive steel pin on a vibrating diaphragm, or membrane, then carefully dipping just the tip of the pin into a cup of electrically conductive acid solution (hence the famous spill). When speaking against the diaphragm, the variations in air pressure caused by the sound waves would make the diaphragm vibrate slightly. The small vertical displacements would be proportional to the amplitude of the sound wave, causing the steel pin

Bell's telephone transmitter (replica). Courtesy of Spark Museum

to move up and down atop the meniscus of the liquid acid in increments proportional to the air pressure variations caused by the sound wave. These subtle variations would cause the electrical resistance of the steel pin and acid to change accordingly, converting the original sound signal into an equivalent, undulating electric current. On the receiving end, the combination of an electromagnet and another connected diaphragm could be used to reconstruct the original vibration of the sound wave as the undulating current flowed through the electromagnet.

Edison's microphone was based on the same principle of variable resistance for sound conversion, but it was implemented in a far superior way. The Edison microphone consisted of an electrical resistor made of fine-meshed carbon granules compressed into the shape of a button. On one broad side of the carbon button was a membrane. Vibrations in air pressure induced by sound waves impinge on the membrane to cause the density of the carbon granules in the button to change in minute increments. This variation in density induced corresponding changes in the electrical resistance of the granulated carbon particles, which led to the modulation of the electrical current passing through them. This solid-state microphone was a revolutionary invention: highly sensitive, compact, simple, reliable, and

Bell's telephone receiver (replica). Courtesy of Spark Museum

inexpensive. It was safe, too—there would be no acid spills with Edison's button microphone. Edison's device also boasted some ability to amplify the electrical signal induced by the sound wave, and as a result, the sound quality produced was loud and clear. Though nobody could have known it, Edison's carbon granule microphone would remain in use in telephone sets for over the next hundred years, until the mid-1970s.

The leap from the telegraph to the telephone represented a great stride in electromagnetic technology. The messages transmitted by original telegraph were digital, either "on" or "off." Telephone signals, on the other hand, are continuous and undulating. This type of signal mimics, or is "analogous" to, the actual shape and form of the stimuli to human sensory perceptions from the natural environment, such as sight and sound. Thus, these continuously varying signals became known as "analog" signals.

The analog telephone quickly grew to be far dearer to the public's heart than the digital telegraph, mainly because it was both so personal and so easy to use. Telephones offered to every man, woman, and child the closeness of natural speech, as opposed to the cold, coded messages of the telegraph. What's more, people could use the telephone whether at home or at work; they didn't have to go to a special telegraph office. Finally, once connected, both parties could converse at the same time, just as if they were face to face having an intimate chat. Western Union initially refused to buy the telephone patent from Hubbard because the company's technical experts assigned to evaluate the invention failed to consider these important factors. But then, understanding the needs and wants of the public is a difficult feat, and it's long been true that technologists and business visionaries often have trouble understanding the needs and capabilities of each other's camps.

After Gray was robbed of his due, he retreated from the field. But two other successful inventors, intrigued by the connection between sound and electric current, continued to do work: Emile Berliner, inventor of the electromagnetic sound-to-current transducer, and of course, the ever-impressive Thomas Alva Edison.

William Orton's special request for a high performance microphone brought Edison into the field of linking electricity with sound. After creating the carbon-button microphone, Edison unveiled his next major invention in 1878: the phonograph. Adapting Gray's principle, Edison attached a needle pointed at a revolving wax cylinder to a membrane stretched across the exit of a cone. When changes in air pressure created by sound waves moving through the cone made the membrane vibrate—the same sound input technology used in the microphone—the needle attached to the membrane etched grooves of differing depths onto a rotating cylinder. When another needle with a softer tip attached to an

Early Edison Phonograph. Courtesy of Spark Museum

electromagnet revolved across those imprinted grooves, it could reproduce and replay the sound waves that had originally created them. In a stroke, Edison invented both the world's first sound recorder and the record player! The general public was totally mesmerized by the idea of capturing sound and then playing it back. To most people, it was simply a miracle.

In his days as a telegraph boy, Edison noticed that skilled operators could simply listen to the clicks of a receiving machine and immediately decode the embedded message. Edison's original intent was to sell his phonograph to telegraph companies to allow them to record the sounds, or content, of incoming telegrams. But telegraph companies did not see much need for this application. To Edison's great surprise, however, he found that the general public became fascinated with idea of recording and playing back music. Seizing the opportunity, he repurposed the device and established the Edison Phonograph Company, a business that exclusively sold phonographs and cylinders with prerecorded music. Edison could record about twenty minutes of music on each cylinder, making the device practical for music playback. The first song to be commercially recorded and produced was "Mary Had a Little Lamb," which ran for six minutes. This marked the first foray of electromagnetic technology into the entertainment market. By the end of the 1880s, bars and taverns in big cities began to specifically equip themselves with phonographs. Customers could walk in, have a drink, and enjoy different music selections. In time, it became a necessity for public houses to offer music to remain competitive.

For his part, Berliner, the man who'd sold his transducer to Bell, immediately recognized the genius in Edison's phonograph and, like any good inventor, set about improving upon it. In 1887, Berliner started his own company and introduced flat vinyl recording discs that could be inexpensively reproduced with a mold, forcing Edison to abandon his cylindrical design. Though the phonograph was a great success, the Edison Phonograph Company, faced with nimbler competition, eventually failed in the 1920s—though this wasn't due exclusively to technological factors. The company's main shortcoming was that Edison himself insisted upon selecting the music to be released, and his personal preferences were not terribly popular with the buying public! For example, Edison preferred music with simple melodies and basic harmonics, and completely missed the growing popularity of jazz and blues. Berliner, however, opened the floodgates to the free release of a variety of selections, and his company eventually evolved into the famed recording company Deutsche Grammophon.

A final word on digital and analog technologies: despite the relative ease with which Edison was able to create the phonograph once the groundwork had been laid, the difficulties involved in the recording, conversion, and transmission of analog signals are typically much more cumbersome than with their digital counterparts. For example, in telegraphy, the attenuation and distortion of digital pulses during long distance transmission can be addressed by using a relay. No such solution existed for the analog signals carried by the telephone. Improving analog signals would require higher quality amplifiers; however, such technology did not exist at the time. As a result, telephone conversations could span only short distances before the voices began to fade or distort beyond recognition. Similarly the volume of music played from a phonograph was quite low even with a large acoustic horn amplifier. New technologies were sorely needed to overcome this formidable barrier and extend the impact of analog signals—technologies that would not come to be invented for another generation.

4

WIRELESS TELEGRAPHY

By the turn of the twentieth century, the telephone was in widespread use. Bell Telephone had close to one million subscribers. The telegraph, however, remained king for all long distance communication needs, and the bellwether for all of this—poles laden with line after line of electrical wiring—covered the land for miles in all directions.

When Morse erected the telegraph line in 1843, it marked the first time that poles designed specifically to support electrical wires had been raised. Morse probably never could have imagined that just a generation later, such poles would seem like clustering stands of urban forest choking the views of the skies. Finally, ordinary people had at last completely accepted electricity and electromagnetism. But all the wiring needed to support the implementations of these principles quickly started to become a nuisance. The scientific foundation laid by Faraday and Maxwell had proven sufficient to support the development and promotion of a wide variety of applied technologies; however, there remained one more important field of Maxwell's theory that had not been explored, one that could perhaps relieve the pressure of all these cables and poles: electromagnetic waves.

HERTZ AND ELECTROMAGNETIC WAVES

In the system of equations Maxwell published in 1873, he predicted a class of electromagnetic waves flying about the atmosphere at the speed of light. Light visible to the naked eye was just a small subset of those waves. There also existed other electromagnetic waves with wavelengths outside the range of what humans could sense.

The man who proved the existence of Maxwell's electromagnetic waves was a young German professor named Heinrich Hertz. In 1888, after eight painstaking years of theoretical research coupled with clever experimentation, Hertz was able to prove the existence of electromagnetic waves with properties exactly in line with Maxwell's predictions. These "Hertzian waves" were just like visible light that humans could detect with their own eyes, but with wavelengths up to a million or more times longer. Hertz's electromagnetic waves could even be reflected by metal plates, just like visual light reflected by a mirror.

To generate the waves that proved Maxwell's theories, Hertz first aligned two brass rods of a specific length, with a small air gap between the two. At the end of each rod, he attached a spherical capacitor. He then placed a high-voltage source across the air gap and controlled it with a switch. As the switch was turned on, the high voltage source would quickly charge up the capacitors, creating a strong electric field in the gap. When the strength of the electrical

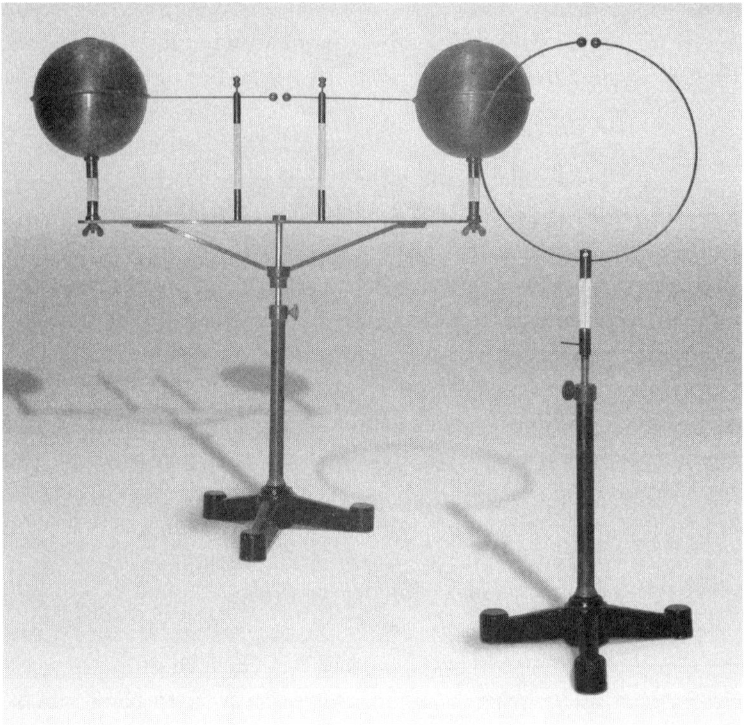

Transmitting and receiving antenna in Hertz's experiment (replica). Courtesy of Spark Museum

field reached a critical level, sparks began to shoot across the gap as the air in between became ionized, forming a conductive path. Electrical charges stored in the end capacitors began to slush back and forth between the two spheres to form an oscillating electrical current, the resonant frequency of which was determined by the capacity of the spheres and the inductance of the rods and the high-voltage source connected across the gap.

Maxwell's theory perfectly and correctly predicted that part of the energy in the oscillating current would be converted into electromagnetic waves, splitting away from the copper rods and radiating into space with the speed of light and a frequency equal to the frequency of the oscillating current. This phenomenon was like to a stone thrown into a lake, with ripples spreading out on the water surface from the point of impact. It could also be analogized as a gong being struck by a mallet, resulting in pressurized waves of sound generated by the vibration of the gong spreading out in all directions. To support propagation of these waves, the ripples need water and the sound waves need air; however, no medium is needed for electromagnetic waves. They can propagate even in the absolute void of empty space.

To detect the existence of this electromagnetic wave, Hertz placed a copper ring near his two rods and capacitors. The ring was specially built to have a diameter of a resonant size relative to the wavelength of the electromagnetic wave to be detected. The ring itself also had an extremely small adjustable air gap, on the order of a fraction of a millimeter. When electromagnetic waves passed through the ring, the ring would intercept part of the energy and produce a weak, resonant electrical current. This led to the production of faint sparks in the narrow gap, which could be detected with the naked eye.

In Hertz's original "spark gap" experiment, the electromagnetic waves that he produced and detected had a wavelength of approximately one meter, or a frequency of 300 MHz (300 million cycles every second), which was consistent with his design. Hertz's work proved to be among history's greatest scientific experiments. It proved that the energy of electrical current could be converted into hitherto unknown electromagnetic waves and vice versa. The rod and the ring in Hertz's experiment served as a transducer between electrical current and electromagnetic waves. Today, we recognize this transducer by a far more common name: an antenna.

MARCONI AND THE WIRELESS

Hertz's experimental results served to validate all aspects of Maxwell's theory and solidify the scientific foundation of electromagnetism. After Hertz proved the existence of electromagnetic waves, many creative minds began to explore their practical applications. Britain's Oliver Lodge proposed using electromagnetic waves in lighthouses, since the Hertzian waves (later named radio waves) could better penetrate the dense fog over water than the flickering light of arc lamps. Other famous pioneers, including Nikola Tesla of Serbia (and later America), Alexander Popov of Russia, and Jagadish Chandra Bose of India, all became interested in finding different applications for electromagnetic waves. But without a doubt, the Christopher Columbus of applying electromagnetic waves to wireless communication was an Italian, Guglielmo Marconi.

Marconi was born into a wealthy family in 1874. Like Galvani of "animal electricity" fame, he was from the vicinity of Bologna, in northeastern Italy. The boy showed aptitude and interest in the field of electromagnetic science from very early on. At the young age of twelve, he was reputed to have successfully reproduced Franklin's kite-flying experiment. In his early twenties, Marconi repeated Hertz's "spark gap" transmitter/receiver experiment, and he

Guglielmo Marconi. Science Source

even went so far as to improve Hertz's signal detection method by employing Oliver Lodge's new "coherer" technology, in which tubes of tiny metal rods suspended in liquid were inserted inside the gap of a receiver ring to detect the presence of radio waves, rather than relying on human eyesight.

Marconi's interest in experimenting with electromagnetic waves was different from Hertz, who was a scientist motivated purely by seeking knowledge. Marconi, however, had a bent toward invention, and he was deeply interested in the potential commercial applications of electromagnetic waves. This business acumen may have come from his mother, whose family came from a line of very successful whiskey distillery owners in Ireland. Regardless, it was Marconi who recognized that electromagnetic waves could potentially be used in communication, sending signals without the need for the physical wires employed by both telegraph and telephone—in other words, that signals could be transmitted and received with "wireless" technology. Marconi also realized that, since Hertz's waves could be turned on and off, a digital messaging method like Morse code was ideal for implementation in wireless communications.

With the assistance of his younger brother, Marconi converted his attic into an electromagnetics lab. In short order, he discovered that by increasing the spacing of the spark-gap in the transmitter, scaling up the voltage of the power source, and raising the height of his antenna, he could transmit electromagnetic signals over longer and longer distances. Soon, he demonstrated that he could use wireless telegraphy to communicate over a distance of three kilometers, employing an antenna attached to a high-flying kite.

At the outset, Marconi's father was very disappointed by his son's indulgence in playing with gadgets and not dedicating himself to the pursuit of an education in literature and business. However, in time, he was won over by Guglielmo's passion and success, and the elder Marconi finally came to support his son's dream to develop wireless communications. In 1896, Marconi, with encouragement from his father, submitted a plan to the Italian government seeking financial support to develop the technology. When the Ministry of Posts and Telecommunications replied politely that it held no great interest in this technology, Guglielmo, with the support of his parents, set off for London to seek out investors and realize his dream. To create a new technology-based business at the cusp of the twentieth century, London, like today's Silicon Valley, was the right place with the right culture and the necessary resources.

With the young and handsome Marconi's dapper appearance, refinement, and grace, he presented a highly polished image among Britain's upper-crust society. His mother's family of distillery owners, very successful Londoners themselves, introduced Marconi to the engineer-in-chief of the British Post

Office, Sir William Preece, and the two men hit it off immediately. Preece had much influence in the British telecommunication establishment, and he was also interested in developing wireless telegraphy based on Faraday's induction principle. Preece would go on to open doors to many opportunities for Marconi to demonstrate his technology to influential members of the public and the press.

In a very short time, Marconi became the talk of London, and with the help of his mother's family, he raised £100,000 (approximately £8.5 million in equivalent purchasing power in 2010) to establish the world's first wireless telegraph company. This sum was a significant amount of money at the time, and Marconi did not spend it in vain. He was a born business genius, and he accurately saw the unique properties of and potential markets for wireless telegraphy. He was also well versed in history and the importance of intellectual property rights, and he applied for and was granted a large number of patents regarding wireless telegraphy, safely securing his future against competitors. Marconi also avoided competing directly with the solidly established wire-based telegraph and telephone businesses; rather, he concentrated on high-valued niche markets where wire-based telegraphy and telephony could not function, such as ship-to-shore communications.

Marconi was deft at using the media to promote both his technology and his name. In 1898, he set up wireless communications terminals on Queen Victoria's royal yacht and at her multiple country homes. News reports on the brief wireless messages sent between the Queen and her son Edward, the Prince of Wales, quickly became the top stories of the day. Marconi's wireless telegraph was also used to report on the events of a famous international yachting race as they unfolded. The location of the race was remote and the nearby land was hilly and without any existing telegraph or telephone infrastructure, so there had never before been up-to-the-minute reporting of the race's events. Marconi and his wireless technology were able to change that. During this process, he also demonstrated for the first time that wireless telegraphy could be used to communicate between two points without the benefit of a direct line of sight.

In 1899, the French government provided Marconi a grant to establish a wireless link across the English Channel. The thirty-mile, open-water connection proved to be a quick success, and it inspired Marconi to begin thinking about addressing a much grander challenge.

CROSSING THE OCEAN BLUE

In the same year he built the cross-Channel wireless connection, Marconi made his first visit to America. The purpose of his visit was to report onsite at another yachting race, the America's Cup, which was to be held in the waters off the coast of New Jersey and New York. Marconi was already a household name in America, and when he arrived stateside, he received a hero's welcome. But while Marconi enjoyed his time in America, when he returned to Europe, he began to wonder if there was more he could achieve. He was able to send messages across the English Channel—perhaps his electromagnetic waves could also be transmitted all the way across the Atlantic Ocean?

Compared to the thirty miles across the English Channel, the 2,000 miles across the Atlantic Ocean represented an enormous leap, not only in terms of physical distance, but also in terms of technological requirements. For a wireless signal to travel 2,000 miles would require enormous signal transmission power, not to mention the fact that the antenna would have to be incredibly large and tall. Furthermore, there were still many fundamental scientific issues that remained unclear. For example, the earth is round—could electromagnetic waves sent from Europe even reach North America? Or would they act like a straight beam of light and shoot out of the earth's atmosphere, directly into space? His technical team conducted a preliminary engineering analysis of the feasibility of transmitting and receiving a wireless signal across the Atlantic Ocean, and concluded that though technically challenging, intercontinental wireless telegraphy should nonetheless be possible. Even though uncertainties abounded, Marconi, being a courageous visionary, decided to launch the project in 1899. After all, countless unexpected perils and complications befell the first transatlantic telegraph cable, but that project succeeded in the end.

Learning from history's mistakes, Marconi recruited the famed British professor John Ambrose Fleming to help design the powerful wireless transmitting station on the English coast. Professor Fleming was extremely knowledgeable about designing electrical systems, and at one time he was an advisor to Edison. Marconi also shrouded his project under strict secrecy, because he knew that, should he fail, it would be a blemish on both his reputation and that of the Marconi Wireless Telegraph Company. He worked with great boldness and vigor, but in truth, he did not have total confidence that he would be successful, so he remained extremely meticulous and cautious.

Marconi built his transmitting station at Poldhu, near Cornwall on the southwest coast of England. After long and careful consideration, the receiving station was chosen to be constructed at St. John's, in Newfoundland, Canada. There were two reasons for the choice of this location: first, Newfoundland was closer to Europe; Poldhu and St. John's were just 1,800 miles apart, slightly short of Marconi's original approximation of 2,000 miles. Secondly, the site was remote, which allowed Marconi to keep the project away from the attention of the news media.

Marconi's basic principle of electromagnetic wave generation was similar to that in Hertz's original experiment, but the scale of the wave generator at Poldhu was far greater, an incredible show of brute force. It was powered by a 25-kilowatt electrical generator with an astoundingly high output voltage of 20,000 volts, applied across a spark gap five centimeters wide. The charge storage capacitors, equivalent to the copper spheres at the end of the rods in Hertz's original experiment, were made of multiple large, parallel plates linked together, and the height of each plate was more than that of ten men. The transmitting antenna, too, was enormous. It was made of wires supported over twenty poles arranged in a circle, and each pole was over 200 feet tall. The system would generate an electromagnetic wave with a frequency of 500 kHz (kilohertz), or a wavelength of 600 meters. During operation of this system, the sparks caused enormous thundering sounds, and the very ground trembled with each firing.

After almost two years of strenuous development efforts and countless failed attempts, on December 12, 1901, a bitterly cold winter day in Newfoundland, Marconi listened to an earphone modified from a telephone receiver and heard a faint repetition of the signal "di.di.di." or "···", the Morse code representation of the letter "S." This signal was sent periodically every hour by the massive transmitting station in Poldhu 1,800 miles away, and it was received by an antenna mounted on a large kite—just as he'd used in his early experiments in his youth—flying at over 300 feet. Marconi appeared to have proved his concept and realized his dream: wireless signals were successfully sent and received across the great ocean.

The news of this monumental achievement spread fast and wide throughout the world. Less than five weeks after his breakthrough, on January 13, 1902, Marconi was honored at the annual meeting of the American Institute of Electrical Engineers in the Waldorf Astoria Hotel in New York City. Many famous guests were present, including Alexander Graham Bell. Thomas Edison, though not in attendance, sent a congratulatory telegram, jibing, "That fellow's work puts him in the same class as me!"

Marconi remained extremely modest at the banquet, with an air almost so quiet as to be mistaken for snobbishness. The truth was, he was bothered by the idle chatter and cutting words of his naysayers, some of whom openly spoke of their disbelief of Marconi's accomplishment right outside the banquet hall. The signals received at the Newfoundland station, it seemed, had been extremely faint. Amidst all the background hissing noise, they faded in and out, emerging and disappearing, and Marconi could not be completely sure whether they had actually been received at all. Perhaps the dots of the "S" were simply blips of random electromagnetic radiation careening around the atmosphere? And if that were the case, what would that say about his experiment? While still enjoying toasts and laudatory claps on the back, Marconi began to wonder how best to continue with his research and deliver the world solid proof as quickly as possible.

After the banquet, Marconi hurried back to England on an ocean liner. While onboard, he mapped out his plan for the next experiment. Upon arrival in Southampton, Marconi arranged to have another ocean liner, the SS *Philadelphia*, immediately fitted out with a huge receiving antenna and with the most sensitive receiver available. After two weeks of intense preparation and testing, Marconi boarded the ship and headed back to New York again, having given instructions to the Poldhu station to transmit every day at regular times a special set of messages, including the "···" or "S" signal. Every day, Marconi sat in the cabin with the ship's captain—a witness—and waited for the signals to be received as the ship traveled farther and farther west. The message was clearly received at a distance of 500 miles and then again at 1,000 miles. At 1,500 miles, some message information began to degrade, but the "S" signal, though weaker, was still distinguishable. Even when the ship docked in New York—some 2,099 miles from Poldhu—the "S" signal could still be clearly discerned. Thus, Marconi was able to eliminate all doubt: he had successfully sent and received wireless signals across the Atlantic. It took another five years of technological refinement before Marconi was able, in 1907, to announce the inauguration of commercial wireless telegraphy service across the Atlantic Ocean.

After his initial success with the transatlantic experiment, Marconi continued to study the properties of wireless signal transmission phenomena in the atmosphere. He observed that the level of background interference was typically higher in the daytime than at nighttime, and he postulated correctly that this was due to a solar radiation effect. Marconi also tried to rationalize one of the matters that originally concerned him before he launched the transatlantic project: why wireless signals sent from Europe could indeed reach North America rather than exiting into space as a straight beam of light. His explanation was that the earth's magnetic field somehow played a role in bending the wireless signal, allowing it to propagate along the surface curvature of the earth. Though this was a sensible speculation, in this case, history would prove him wrong. It was not until 1927, with the identification of the ionosphere—an outer layer of the earth's atmosphere 50 to 300 miles from the earth's surface, which contains ionized gases that refract radio waves and bounce wireless signals back toward earth—that the true reason could be understood.

In recognition of his contribution to wireless telegraphy, Marconi was awarded the Nobel Prize in Physics in 1909, together with German scientist Karl Ferdinand Braun. He was probably the only Nobel laureate in physics who did not formally graduate from a university, but in light of his extraordinary accomplishments, he certainly deserves the honor.

In the wake of Marconi's successful trials, many big ocean liners began to install long distance wireless telegraphy equipment onboard. In building up this business, Marconi pursued a new business model: the company would not sell any wireless equipment directly to customers. Instead, it leased the equipment, providing all services required, including design, installation, maintenance, and operation. Thus, all wireless telegraph operators on board ships would be Marconi company employees, and Marconi could charge a steep price for this total turnkey solution.

The job of a Marconi Company operator was a skilled and somewhat prestigious position, just like that of an elite telegraph operator had been in previous generations. When the *Titanic* sank in 1912, there were two Marconi Company wireless operators on board. After the ship's fatal collision with the iceberg, the operators sent off many distress signals for help. As it happened, there was another ocean liner in the vicinity, one that also had Marconi communications services onboard. This ship received the *Titanic's* signal, and within four hours, it was able to arrive at the scene and provide emergency assistance, saving over 700 lives. One of the *Titanic's* wireless operators drowned, sending and receiving telegrams in the midst of the disaster, right to the very end. If there had been no wireless communication capabilities, the fatalities of the shipwreck, bad as they were, would have been far worse. Marconi emerged as a hero in the wake of the *Titanic* tragedy. Businesswise, he also came out as a winner, as new rules were established that required all ships above a certain tonnage to have wireless communication capability onboard. This proved a financial bonanza to the Marconi Company.

In the years that followed, Marconi Wireless Telegraphy Company continued to grow. Unfortunately, a number of challenges, both professional and personal—the technological innovations of Reginald Fessenden and Lee De Forest; the car accident that injured Marconi's spine and claimed one of his eyes—slowed down Marconi's involvement in business and dampened his entrepreneurial spirit. In 1917, Marconi moved back to his birthplace in Italy and began to turn his interests toward politics. He received the honorific title of marquis from the Italian government, and later, he developed a close personal relationship with Mussolini. Eventually, Marconi's nationalism led to his becoming a National Fascist party member, and when Marconi died in 1937, Mussolini was at his bedside.

Despite the sad political affiliations he claimed late in his life, Marconi was an outstanding technologist and entrepreneur. With his exceptional vision, ability, dedication, courage, and business acumen, he pioneered and built the first wireless communications industry. He was instrumental in bridging the gap between the scientific knowledge of Maxwell and Hertz and the needs of the society at large, and we undoubtedly owe the countless benefits of electromagnetic waves we enjoy today—including satellite radio and TV, mobile phones and Wi-Fi—to the boy from Bologna, Guglielmo Marconi.

5

LIGHTING AND ELECTRIFICATION

Throughout the final years of the nineteenth century, the newly discovered principles of electricity and magnetism found many uses. With the telegraph, the telephone, and wireless telegraphy, the obstacles of distance and time in communication were effectively eradicated. However, communications was not the only field in which electromagnetic technology was making an impact. The creation and popularization of electrification—electrical power and lighting systems—represented a still more fundamental and profound change to the way humans lived.

ELECTRICAL LIGHTING SYSTEMS

The first electric lamp was created by Volta, who demonstrated it to Napoleon when he heated a wire with his battery. Running electrical current through a thin metal filament caused the metal to heat up, as electrical energy was converted to heat energy by the electrical resistance of the metal itself. At first, this heat caused the filament to glow a dark red, but as current increased, the filament temperature would continue to rise and the emitted light would turn bright white. Unlike the arc lamp first demonstrated by Davy in 1807—which used two carbon rods sharpened at the tips, across which high-voltage electricity would spark and jump in powerful, glaring bursts—the light emitted from the heated filament, or the "incandescent lamp," was soft, stable, and pleasing to the eye. It was suitable for use both indoors and out. For hundreds of years, it had been a dream for human beings to find an effective and safe way to enjoy sustained light at nighttime or in the dark. Burning wood, candles, and oil lamps had been used throughout history. In the mid-1800s, gas lamp lighting, a major improvement, became popular. But might electromagnetics hold the key to a better and safer lighting method?

In 1878, Thomas Alva Edison, who had already enjoyed tremendous success with his major entrepreneurial inventions in telegraph, telephone, and phonograph, decided to tackle the problem of using electricity to create and provide lighting. Edison recognized that even though crude incandescent lighting had been demonstrated, there still remained two fundamental problems to be resolved before it could become widely adopted. First, incandescent lights consumed a large amount of electrical energy. Using batteries as their power source would be exorbitantly expensive and not at all practical. Thus, a method for continuously and cheaply generating a plentiful amount of electric power, and some infrastructure for effectively delivering that power to homes, would be a necessary pre-condition to any practical solution to large scale adoption of electric lighting. This was a formidable challenge in total systems engineering, and it was beyond the capability of any one man—even the brilliant Edison. The second problem was that of the light source itself. At the time, experimental lights as demonstrated by numerous inventors around the world generally

Edison in 1878 working on his phonograph. Getty Images

burned out within hours. To make a lighting system practical and affordable, Edison knew he would have to extend the functional life span of the incandescent light to at least several hundred hours to convince the general public that its adoption merited the expenditure.

As early as 1878, just after he'd invented the carbon-button microphone, Edison foresaw that the solution for cheap and plentiful electrical power might soon be possible. In the late 1870s, European technologists achieved a major breakthrough in electrical power generation technology. With the fundamental power generation problem—if not the matter of its distribution—solved, Edison correctly surmised that the time to launch his lighting system project had finally arrived. His plan was two-pronged: he would extend the useful lifetime of incandescent lights, and at the same time, he would commence developing an electrical power generation and distribution network to provide affordable power to the users. This total system solution oriented thinking by Edison was far beyond what other inventors at the time were able to envision and comprehend, and this offered Edison a strategic advantage. To launch the effort, Edison raised $150,000 and kicked off his project with great fanfare. His initial market entry point was very specific: to replace all the gas street lamps in New York City—and their supporting infrastructure—with an integrated, electricity-based lighting system.

Edison's early experiments were disheartening. As with countless other inventors, his electric filaments lasted only a few hours before burning out. However, Edison remained persistent. Using a trial and error methodology, Edison and his team systematically experimented with hundreds of different materials, searching for a robust solution. In time, they discovered that high-resistance carbon-fiber filaments, crafted by carbonizing long-fiber cotton or bamboo, offered the best solution. These filaments were relatively long lasting, glowed white quickly, and were inexpensive. With further refinement, including cocooning these filaments in vacuum-sealed enclosures (that is, a glass bulb), the average useful life of these fibers was extended to several hundred hours, with the best reaching upwards of one thousand. His idea to utilize a brute-force approach based on the systematic use of the trial-and-error method, commonly referred to as the "Edisonian Approach," was ideal under the circumstances: there was little effective knowledge available that might have allowed him to attack this technological problem using theoretical calculations alone. Edison wasn't the first inventor to think of placing his filament in a sealed glass enclosure—Humphry Davy had conceived of the idea some seven decades earlier in his carbon-tip spark experiment—but his model performed far better than any previous iteration.

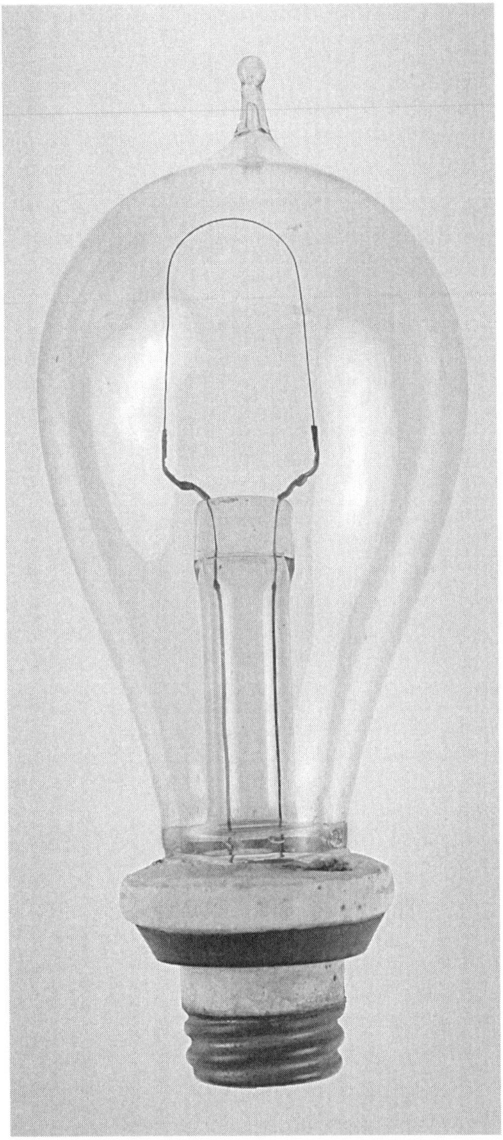
Original incandescent light bulb used in Menlo Park demonstration. Courtesy of Spark Museum

In parallel with the light bulb project, Edison and his team designed and built a model system for electrical power generation and distribution. Their power system included a steam engine-driven generator, transmission cables, switches, fuses, and even power meters, the last of which were essential to providing billing information: Edison planned to charge users not a flat fee, but in increments relative to the amount of power used.

By creating a practical light bulb and basic power distribution system, Edison and his team proved their technical aptitude and business insight. However, in reaching these two milestones, Edison had spent much of the money he initially raised. In order to realize his vision of building a real, commercially viable electrical lighting system, Edison would have to leave the laboratory and attract more funding.

Many of the major capitalists of the day, such as J. P. Morgan and Cornelius Vanderbilt, had invested heavily in the telegraph business with Western Union. As the telegraph company and its backers found themselves in the late 1870s locked in a battle over the telephone—a technology on which many were still taking a wait-and-see stance—financiers were interested but understandably reticent with respect to Edison's electric lighting system. Nonetheless, Edison was convinced that, with the right demonstration, he could convert even the most skeptical potential patron. On New Year's Eve, 1879, Edison invited a large, carefully selected crowd of guests, including news reporters and officials from the New York City municipal government, to his laboratory in Menlo Park, New Jersey. It had snowed earlier in the day, and when a specially commissioned train from Manhattan unloaded outside Edison's lab, its passengers were greeted by Edison, standing before a muted, resplendent landscape of pure white.

As the guests stood in the growing darkness, unsure of what to expect, Edison turned on the generator and, with a flourish, firmly pressed a single switch. Instantaneously, the grounds were lit with the soft and comfortable glow of 290 electric bulbs shining divinely above the blanket of freshly fallen snow. The crowd was agog, but Edison was not yet finished. He fiddled with another set of switches, and row after row of lights blinked on and off. This was the world's first string of electric Christmas lights! No other existing light source could be controlled so easily, and even the wizened official in charge of managing New York City's gas lamps watched in open-mouthed amazement. The next day, January 1, 1880, all of New York's newspapers ran Edison's grand accomplishment as their cover story, proclaiming with the start of the new decade the commencement of a new era for humanity. No longer could the dark of night curtail man's activities.

The Menlo Park demonstration proved the feasibility of both the incandescent lamp and Edison's power distribution system concept, and the deep-pocketed attendees of the affair were, as expected, immensely impressed. Edison, ever the savvy businessman as well as brainy inventor, took advantage of the interest he'd so personally sparked and proposed to build his first, small-scale power distribution system, including a central power generation station, in a wealthy neighborhood near downtown Manhattan. This distribution system, the first "power grid," would supply electrical power to every lighting subscriber in that neighborhood—one in which, as Edison knew, the splendid Manhattan residence of powerful financier J. P. Morgan stood. After Edison's Menlo Park show, it took very little effort to convince Morgan and other commercial barons like him to invest in the project.

Now armed with ample funds, Edison got down to work. In 1882, the world's first commercial central power station commenced operation. The Pearl Street station employed a 27-ton electrical generator, which, using Edison's improvements to existing European designs, could offer a maximum output of 100 kilowatts of direct current (DC) power at a supply voltage of 110 volts. This was enough electricity to power 1,200 light bulbs—an astronomical number, by the standards of the day—and the residence of the largest of Edison's fifty-nine subscribers, J. P. Morgan himself, thenceforth blazed with light every evening, becoming one of the grand sights of pre-Prohibition era New York City.

With its unqualified success in Manhattan, the business of electrical lighting systems reached its tipping point. Through licensing Edison's technology and patents, many new electrical power plants and distribution networks were constructed across the nation, similar to the buildup of the national telegraph system after Morse's successful Washington-Baltimore demonstration. Within four short years, well over one hundred power plants and associated power distribution systems were put into operation across America. At the same time, J. P. Morgan set in motion the founding of Edison General Electric Company—known as Edison-GE—which specialized in developing and manufacturing key electrical products such as light bulbs and generators. Named a founder, Edison received a significant share of the equity in this company, achieving both fame and fortune as a result of his exceptional vision, courage, and hard work. In 1882, Edison-GE produced 50,000 light bulbs; by 1887, that figure reached over 1,000,000, and Thomas Alva Edison was believed to possess the second-highest name recognition in the world, behind only Queen Victoria. A major new industry was born, and Edison was at its helm—although a thorough examination of history shows that he owed his success to the clever integration and practical improvements he made to the electrical power generation technology developed in Europe several years earlier.

GENERATORS AND MOTORS

It was Michael Faraday who, in 1821, first demonstrated the scientific principles of the electric motor. In his classic experiment, he converted the chemical energy of a battery into electrical energy flowing through liquid mercury. He then transformed that electrical energy into kinetic energy via the motion of a spinning copper rod. This demonstrated the principle of a motor in its most rudimentary form. Ten years later, Faraday demonstrated the feasibility of converting kinetic energy back into electrical energy. By moving a magnet back and forth through a coil of wire and observing that electrical current flowed through the wire as a result. This kinetic-to-electrical power convertor was, at its core, the first electrical power generator.

Even though the basic physical laws governing motors and generators were well understood long before Edison's time, it had long proven extremely difficult to develop practical products. Early motors were like toys: even though the designs were clever, they did not produce enough power to be of practical use. One of the limitations was the fact that for years, batteries were the only available source of power for motors, yet they were clearly too expensive for wide scale implementation. Therefore, from the mid 1840s on, most inventors' attention was shifted to the development of generators, and in the ensuing twenty years, generator design technology progressed steadily.

Over this period of time, the basic designs of these generators began to converge, and they consisted mainly of four parts: the stator, the rotor, the armature, and the commutator. The stator was generally the stationary part of the generator, which supplied the magnetic field necessary to create electricity. The rotor was the rotating shaft, which was mechanically driven by a steam machine or, later, a turbine. A coil of wires, known as armature, was typically attached to the rotor. When the rotor revolved in the magnetic field inside the stator, the mechanical energy from the external power source would be converted to electrical energy in the form of periodic waves of electric current flowing through the armature windings. These varying-amplitude electrical currents were known as alternating current, or AC, which was different from the electrical energy that flowed from sources like batteries in constant, direct current, or DC.

The last part of the generator was the commutator, which transformed the alternating current produced in the armature coil into steady, direct current for output. The commutator was necessary because most of the market demand for electric power at the time was for DC current. The new DC generator, also known as the dynamo, had the market potential to replace the battery in many well-established applications. Yet, the sole market for alternating current at the time was for electric arc lamp lighting, which was much smaller.

Early generators, even successful ones, tended to be highly inefficient at converting kinetic energy into electrical energy. Major breakthroughs would still be needed to make the generator commercially practical. The first such breakthrough was made by German engineer Werner Siemens in 1867. He recognized that one of the primary limitations of the dynamo was that the magnetic field strength in the stator was too low. To solve this problem, he cleverly designed a system that used powerful electromagnets to replace all permanent magnets in the dynamo, thus solving one key engineering challenge to improving conversion efficiency.

The second major breakthrough came from Zenobe Gramme, the son of a Belgian carpenter. He noted an additional, subtle design flaw to the dynamo: the spatial gap between the stator and the armature/rotor was typically too large, significantly weakening the coupling efficiency between the coil and the magnetic field and sapping the dynamo of much of its potential output. However, this problem was not easily corrected, as the revolving armature had to maintain

Zenobe Gramme. SPL/Science Source

a safe distance from the stator to avoid striking it during high-speed rotations.

Through ingenious design and clever use of new materials, Gramme was able in 1871 to develop a much-improved version of the dynamo. His timing was perfect: by the 1870s, there had grown a strong market for cheap, inexhaustible electrical power sources to replace the expensive batteries then used in the telegraph and electroplating industries. As a result, demand for Gramme's dynamo quickly soared.

In 1873, Gramme and his French partner, Hippolyte Fontaine, exhibited their dynamo products in Vienna. During one of their demonstrations, Fontaine mistakenly connected two dynamos that stood far apart from one another via an electric cable. During a demonstration, one of these dynamos was mechanically coupled to a steam engine to generate electrical power. Unexpectedly and to everyone's total amazement, the other connected dynamo started to turn. This was a fortuitous mistake and a historic moment! Not only was this the first demonstration of a practical motor, it was also the first time that a considerable amount of electrical power was transmitted over any sizeable distance. This twin realization—that a dynamo's electrical output could be sent to a remote location through a wire, and that running a high-efficiency dynamo in reverse would essentially turn it into a high performance DC-powered motor—was an enormous breakthrough. Serendipitously, Gramme brought about one of the key technical innovations that started the Second Industrial Revolution!

Despite his early success, Gramme worked hard to continuously improve the dynamo design and expand its market. Another modification he made—and an important one—was to remove the commutator from the dynamo altogether, essentially converting the DC generator back into a simple alternating current generator, or *alternator*. With his alternator, Gramme was able to address what was then a growing market for electric arc lamps. In 1874, nearly eight years before Edison's incandescent light bulbs flooded the Manhattan night, Gramme illuminated the exterior of the Paris Opera House with a brilliant array of arc lamps, using his alternator to set the skyline aflame.

News of these breakthroughs in dynamo and motor technology in Europe led Edison to sense that the timing was right to develop a practical electrical lighting system. After finishing developing the carbon-button microphone and the phonograph, Edison threw four years of effort into developing the light bulb and a central lighting system, for which he achieved unprecedented success. After his successful 1882 opening of a central power station system in Manhattan, power plant construction spread all over the country like a firestorm, and it was not long before electrical lighting became both widely accepted and progressively more affordable, leading to an abundance of moderately priced electrical power running straight into people's homes.

Edison's power distribution system was based on low-voltage (110-volt) direct current. He viewed AC technology as unnecessary and potentially dangerous, eschewing its use in any application beyond outdoor arc lamp flood lighting. In further deference to safety concerns, Edison spared no expense in burying his electrical wires underground, away from possibly hazardous direct human contact.

Edison's electric power and lighting system worked well, except for one shortcoming. When transmitting low-voltage, direct current through normal copper wires, the loss of electrical power due to the intrinsic resistance of the wire itself was moderately high. Over longer distances, the cumulative losses grew so great as to be prohibitive. To alleviate the problem, the only viable solution was to use larger-diameter copper wires to reduce line resistance, which led to significantly higher cost. Due to this fundamental limitation, Edison's DC electrical power grid could only effectively supply subscribers within a radius of about 1.5 miles from a power station, and Edison was forced to resign himself to the fact that his DC power plants would have to be small-scale and located in the heart of densely-populated areas.

Since Edison's DC power system was the only available technology and this specific technical limitation did not directly impact the users' appetite for light, this shortcoming didn't stymie the growth of the electrical lighting industry. At Edison-GE, orders for dynamos and light bulbs grew exponentially, and the company's profits surged in lockstep. However, by 1887, Edison began to notice, to his great surprise, that orders were beginning to taper off, and many of his potential clients were instead buying from a little-known but rapidly growing new company from Pittsburgh called Westinghouse Electric. More surprisingly, Westinghouse, unlike Edison-GE, was using not DC but alternating current—AC for lighting.

THE AC-DC WAR

Throughout his life, George Westinghouse was both a hands-on inventor and an industrialist. In his youth, he invented a railroad signal light system and later an airbrake technology used by trains. By 1885, he had already created four successful companies and made himself a millionaire many times over, but he always remained on the lookout for new business opportunities.

Early in the 1880s, one of his business representatives noticed a news report in Great Britain about using high-voltage alternating current to minimize power loss in electrical wires. When Westinghouse heard this, he realized that another grand opportunity might have arrived. Alternating current had a unique property that direct current did not: its voltage could be readily "stepped up" or "stepped down" by using a transformer, one of Faraday's great inventions. Furthermore, high-voltage alternating current transmitted by electric wires lost far less energy than low-voltage direct current, allowing AC electrical power to be sent efficiently over long distances, even dozens of miles. Of course, all of Edison's power plants—far and away the majority in the United States—output direct current exclusively, limiting their range and size. Westinghouse envisioned a competing power source: massive, AC power stations built far away from subscribers in areas with cheap land prices. The scale of these stations could be made large, in order to satisfy subscriber demand over a broad area and create an economy of scale in operating cost. The AC output power from these remote power stations could be stepped up to very high voltages, then, with very low loss, transmitted over long distances to the

George Westinghouse. Photo Researchers, Inc./ Science Source

vicinity of the subscribers. Once it arrived, the voltage could be stepped down through transformer stations to levels safe for domestic use. Since the incandescent light bulb could be operated with either DC or AC electricity, customers could use either and not notice—or care about—the difference.

When Westinghouse saw this fundamental superiority of alternating current in large-scale applications, he quietly got down to work and prototyped his first AC power plant. He bought an alternator from Siemens and then optimized its performance. Then he bought a transformer from a company in Great Britain and modified it to handle a high power load. Within a few months, Westinghouse was able to integrate the AC alternator with the transformer, and performed the voltage step-up and step-down function with high efficiency. Once he was satisfied with his work, he began producing his own products—AC products—and entered the electrical lighting system market.

With his efficiencies of scale and decreased transmission losses, Westinghouse was able to successfully underbid Edison on most projects. Edison was surprised, expecting more public support for his company, but the public didn't care—they largely didn't care about the difference between direct and alternating currents; the extent of their interest was that Westinghouse's price was lower while the quality of lighting he provided remained identical. By 1887, just five years after Edison-GE had singlehandedly created the electric power industry, Westinghouse had already won twenty-seven large public and private projects to build electric lighting systems, including central power stations.

Edison recognized the seriousness of the threat Westinghouse presented, and his instinctive reaction as a proud, authoritarian technologist and a stubborn businessman was to fight back. He launched a massive public campaign to attack and discredit Westinghouse and AC power, thus kicking off history's famous War of the Currents, also known as the AC-DC War. Edison considered high-voltage power lines strung over urban areas to be extremely dangerous, and he held a press conference to point out the potentially mortal dangers of alternating current. To dramatize the point, Edison had someone catch a stray dog, and in front of everybody, he electrocuted it with high voltage AC current. When this failed to produce the desired public outcry, Edison progressed to still larger animals, electrocuting a horse, and then a zoo elephant that had gone berserk. Still unsatisfied and ignoring advice that he may, perhaps, have gone too far in his grotesque displays, Edison designed a killing machine that employed high voltage AC current to execute criminals. He maliciously dubbed the practice the "Westinghouse-style" execution. Today, we call it the electric chair.

Despite the grisly nature of Edison's demonstrations, Westinghouse's AC business continued to grow at the expense of Edison-GE. He did his utmost to improve the safety of the high-voltage transmission system, depriving Edison of any weaknesses to exploit—indeed, Edison's insights into the initial drawbacks of AC systems only led Westinghouse to ways by which improve his product. In 1889, as part of a counterattack on Edison's cawing about the danger of alternating current, Westinghouse obtained patent rights to the newly developed multiphase AC motor, the first high-efficiency, high-power motor to run on AC power. It was invented by the man who would become Edison's nemesis: Nikola Tesla. With the AC motor added to his portfolio, Westinghouse controlled an entire vertically integrated product family of AC-based electrical equipment. His company's consolidated position became unshakable, even by Edison.

In time, it became clear to everyone that AC technology was the superior choice for power distribution. However, Edison was obstinate, and he continued to desperately attack and slander high-voltage alternating current. At first, he may have been sincerely worried about its dangers, but over time, he became inordinately obsessed. He had, he felt, staked his reputation on this battle, and to him, an admission of AC's superiority was tantamount to an outright surrender. Edison had never lost before; the failure of DC to win the War of the Currents led Edison to become increasingly irrational. This madness of temperament had a negative impact on Edison-GE's business, and major shareholders in the company—including primary investor and controlling shareholder, J. P. Morgan—began to grow increasingly unhappy with Edison and his antics.

John Pierpont (J. P.) Morgan. © Oscar White/CORBIS

In 1891, Morgan traveled to Europe and visited the Siemens Company, the company built by Werner von Siemens to develop the telegraph industry in Europe. By then, the company had introduced an improved power generator and expanded to become the world's largest and most powerful enterprise in the entire electrical industry. In discussions with Siemens officials, Morgan became convinced that the future of the electrical industry was in alternating current and that Edison-GE—and perhaps, Edison himself—was too narrowly focused. Upon his return to America, Morgan quietly set about orchestrating a merger between Edison-GE and the Thompson-Houston Company, which held a number of innovative patents for electric arc lamp lighting systems and produced a considerable portfolio of AC products. J. P. Morgan felt that, despite Edison's railings, diversifying to play in both the AC and DC spheres would best support his company's future and allow him to protect his investment against the encroachment of Westinghouse.

In 1892, Morgan announced the merger of his company and Thompson-Houston, as well as their combined subsequent name change to General Electric Company, or simply GE. Edison's name was excised entirely, and the company presidency, too, was granted to Thompson-Houston's Charles Coffin, a former shoe salesman. Edison was kept in the dark throughout this whole process, learning of his ouster only through rumors that were eventually confirmed by the newspapers. Without any fanfare, Edison was forced off the stage, crushing the self-confidence of this brilliant man.

With Edison out of the picture, the War of the Currents was AC's to win, and a natural battleground was already waiting. The Chicago World's Fair of 1893—also known as the Columbian Exposition—issued a request for proposals to power the 92,000 light bulbs that would illuminate its vast grounds. Westinghouse beat out GE and won the contract. For the Fair, the company developed enormous alternators; the revolving armatures alone weighed twenty-five tons. Westinghouse engineers and Nikola Tesla collaborated to design the specially constructed AC power system, implementing a "multiphase" distribution concept in which two to three AC sources transmitted their energy on a single wire with precise synchronization. This power network proved robust, efficient, and safe; it provided enough electricity to power up to 120,000 incandescent light bulbs and arc lamps, as well as several hundred AC motors. The architecture of the entire grid was flawless, and its design became the de facto standard for electrical power systems in the United States. In fact, Tesla's design remains the industry standard to this day!

In 1895, Westinghouse won another milestone contract, this time to build the unprecedentedly immense Niagara Falls hydropower station. This project called for ten turbine-driven generators to produce 50,000 horsepower of electrical power and, using transformers to increase its voltage to an astounding 22,000 volts, transmit the electricity through power lines to burgeoning Buffalo, New York, some 27 miles away. The Niagara Falls contract served as the decisive victory for alternating current, and from then on, direct current power transmission systems became largely irrelevant. In recent years, however, high-voltage DC technology has found new use in long distance power transmission between large power grids. Still, by the dawn of the 1900s, there was no doubt that the vision of Tesla and Westinghouse—AC power—triumphed over that of Thomas Alva Edison.

IMPACT OF ELECTRIFICATION

Innovation in electrification did not slow while waiting for the War of the Currents to play out. With the advent and burgeoning availability of practical generators and motors, electric-powered creations began to flood cities and factory towns throughout the late 1800s, impacting human lives at all levels and in all industries, including manufacturing, mining, agriculture—and transportation.

The first street trolley was built in Germany in 1876, by none other than the Siemens Company. The trolley was powered by an off-board generator driven by a steam engine, and electric wiring strung in the air above the street surface delivered the electricity to an onboard motor in the car. The design was simple and, with the exception of the unsightly cable line, elegant; the cars were fast and free of noxious exhaust fumes, and they quickly became a favorite of city-dwellers. From Germany, the trolley expanded worldwide, and the resulting boon for Siemens led the company to construct the world's first electric-powered trains, an invention that proved both powerful and timely.

By the end of the nineteenth century, the matter of public transit had become a pressing question in cities throughout the world. The trolley was a welcome addition to cities with modest populations, such as San Francisco and Amsterdam. But large cities with very high population densities needed a higher capacity solution for mass transportation. In order to avoid further crowding on their already packed city streets, the British began in 1863 to build a railway underground. Initially, the London Underground ran exclusively on steam engine-powered trains. The attendant ventilation problems, however, quite literally nearly choked out the rail network in its infancy. It was not until the introduction of nonpolluting electric propulsion technology that the shortcomings of the subway system were solved. By 1890, the London Underground was completely electrified and its transit network greatly expanded in scope. The system was such a success that other major cities began to emulate it. In 1894, Siemens was commissioned to design and build the world's second underground system in Budapest, Hungary. New York City followed shortly thereafter, and almost 120 years later, the wave of electric-powered subway construction in the world's big cities continues to grow, particularly in China.

Like trains, early cars and powered carriages were first driven by steam engines; very rapidly, battery-powered electric motors took over. However, the performance and range of electric cars were severely limited by the constraints of battery technology. Therefore, as the internal combustion engine matured, the personal electrical car began to be phased out in favor of gasoline power. Since 1900, gas-powered cars have outsold electrical vehicles. The problem of running cars on battery power remains to this day a serious technical challenge, but the twenty-first century appears poised to see breakthroughs in technology needed to reverse this hundred-year-old trend. The initial success of Tesla Motors' 2013 Model S sedan, for instance, which cleverly employs over seven thousand Li-Ion batteries, has raised new hope for the future of electric cars.

As with trains, few boats pursued electrical power. Thanks to wind energy and steam turbines, there was little need. However, the unique requirements of underwater vehicles—submarines—altered the picture. When cruising underwater, the lack of available oxygen makes combustion impossible. Thus, the first submarine—built in 1888 by the Spaniard Isaac Peral—utilized large banks of high-capacity storage batteries that powered the vessel while submerged. These batteries had to be charged by onboard generators while the vessel was surfaced. This restriction was overcome when the nuclear-powered submarine was developed, but still, electricity is what makes submarine travel possible. Electricity generated by submarines' onboard nuclear power plants is used not only for propulsion, but also for all life support systems, including production of fresh water and oxygen from seawater. Modern submarines can stay submerged for months at a time, and they truly represent what marvels the innovative application of electrical technology can bring to life.

Aside from lighting and transportation, manufacturing industries also reaped the benefits of electrification. Most factories had been powered nearly exclusively by steam engines since the days of the First Industrial Revolution. Mechanical energy was transmitted to the machines via a complex system of rotors, cogwheels, and belts. This type of mechanical energy transmission system was complicated, bulky, inefficient, and prone to breakdown. But with the emergence of the generator, electrical energy could be directly provided via wires to motors that were integrated with the machinery. In this way, manufacturing efficiency vastly improved, and costs were reduced across nearly all major industries.

The sudden ubiquity of the electric motor ushered in many new applications, which themselves bore still more descendants. Siemens and an American named Elisha Graves Otis were the first to use the motor to invent elevators, making practical the construction of skyscrapers. The motorized electric pump became an indispensable tool for pumping water, which led to the possibility of undertaking drainage projects, building dams and canals, irrigating large tracts of agricultural land, and mining previously unobtainable metals. Finally, electric pumps found application in compressors and other forms of "heat exchangers," which made possible air conditioning and refrigerators. These home appliances followed electric lights into the household and changed our daily lives.

EDISON, TESLA, AND SIEMENS

Any discussion of the powers of electrification warrants a quick recounting of the histories of the three men whose footprints left indelible marks upon the process: Edison, Tesla, and Siemens.

Thomas Alva Edison was born into a poor but upright Ohio family in 1847. From his earliest boyhood, he was bright and studious as well as business-savvy and street-smart, selling newspapers and fruit on passenger trains from the age of twelve. Under the tutoring of a friend's father, he learned Morse code, and at age sixteen, Edison left home to ply his trade as a telegraph operator. From then on, his fortunes would be inexorably intertwined with those of the wire and the telegraph.

Thomas Edison. Photo Researchers, Inc./Science Source

As he was exceptionally capable and skilled, it didn't take long for Edison to make a name for himself as a telegraph operator. In short order, he was assigned to work the busy Boston desk. Although he had little formal education, Edison never stopped trying to improve himself, and he made a point of learning the basics of electricity, chemistry, and mechanical design. Technological innovations fascinated him, and in his early years, he invented a number of improvements to telegraph technology, including the quadruplex telegraph, the automatic telegraph, and the stock ticker-tape machine. These brought him a fortune at a quite young age, and he never forgot that the telegraph was the launching pad to his success—he even nicknamed his first two children Dot and Dash!

In 1876, Edison used his earnings to establish the Menlo Park laboratory, a New Jersey facility devoted exclusively to new technology-based product development work. This was the world's first multidisciplinary industrial research and development organization, and it counted among its staff electrical engineers, chemists, mechanical engineers, and even applied mathematicians. Many big companies, including Western Union, frequently commissioned Edison's laboratory to develop new products and enhancements to their own in-house technologies. In 1877, Western Union's president, William Orton, contracted Edison to design a transducer for converting sound signals into electric current, thus giving birth to the carbon-button microphone and revolutionizing the telephone. Next came the phonograph, which for the first time allowed the human voice to be recorded and replayed. As Edison kept churning out inventions, members of the media took to calling him the Wizard of Menlo Park. Of course, no innovation could match the extravaganza he put on for New Year's Eve, 1879.

When Edison successfully demonstrated electrical lighting on a massive scale, his reputation reached its zenith. However, when AC power challenged his reign—and in less than a decade, dethroned him—Edison found himself falling into disfavor. Edison's own stubbornness and his refusal to accept the will of the masses or others' scientific advances led to his removal from General Electric, a company he had helped found. Subsequently, he refused to ever again engage in electrical research, and in a fit of anger, he sold off his GE shares at a substantial financial loss. He then invested that money in a dubious technology that aimed to enhance the purity of low-grade iron ore. After losing a fortune backing this venture and watching GE shares skyrocket in value, he remained unrepentant. "The most fun," he declared, "was spending that money!" He never became very wealthy again.

In his later years, Edison devoted his energies to developing motion picture technology, with some limited success. When he moved to Florida and semi-retired, he developed a deep and abiding friendship with two younger men: the automobile magnate Henry Ford and the rubber king Harvey Firestone. Both men treated him with great respect and retained him as a special technical consultant. These arrangements allowed Edison to sustain his standard of living. To show his worth, Edison developed an important battery technology based on alkaline metals for Ford to use in electric vehicles. For Firestone, Edison synthesized a new material that could be used as a substitute for natural rubber. In time, Edison also was offered and accepted a special appointment from the U.S. government to help establish the famous Naval Research Laboratory in Washington, DC, which played a major role in such technical breakthroughs as the development of electrical power systems for nuclear submarines.

Edison was a man fueled with an intense, trailblazing spirit. He was also the ultimate workaholic, amassing a total of 1,093 patents throughout his life. Edison was also an extreme pragmatist, and he showed open contempt for empty talks by academics and for overly fancy inventions that lacked commercial applications. His motto was, "Genius is 1 percent inspiration and 99 percent perspiration." Edison died in 1931 at the age of eighty-four, after which the fruits of his life-long labors were placed on permanent exhibition at the Henry Ford Museum in Michigan.

Despite his extreme obstinacy—and the over-developed sense of self-assuredness that led to his defeat in the AC-DC war and the failure of some of his businesses—Edison remained one of the most outstanding contributors and heroes of man's eventful journey toward harnessing the power of the tiny electron to enhance our lives.

Nikola Tesla. Science Source

It's almost impossible to speak of Thomas Alva Edison without mentioning the man who became his adversary: Nikola Tesla. Tesla, a Serbian, was born in 1856 and studied for two years at the Austrian Polytechnic in Graz before bouncing around central Europe, eventually settling in Budapest, Hungary,where he began a career in a then still-burgeoning field: telephony. After two years spent working as an electrical engineer for the National Telephone Company, Tesla joined the Continental Edison Company in its Paris branch office. His talent and intelligence were quickly appreciated by his supervisor, himself previously one of the top managers at Edison's Menlo Park laboratory, who in turn introduced Tesla to Edison. Edison was impressed with the young engineer, and in 1884, the twenty-eight-year-old Tesla was coaxed across the Atlantic to begin work in Edison's office in New York City.

In all his endeavors, Edison always considered himself the absolute and final authority. Tesla, too, had a high opinion of himself, making a clash between the two egos inevitable. Referring to Edison's "ninety-nine percent perspiration" motto, Tesla once said, "If Edison could just think for a bit, then ninety-nine percent of his efforts would be unnecessary." The friction between them came to a head just a year after Tesla relocated to New York.

Despite his outlook—or perhaps because of it—Tesla rose quickly through the ranks in Edison-General Electric. In short order, he found himself charged with improving the design of one of the company's key products: a medium-power dynamo, the DC generator. Edison half-jokingly promised Tesla that if he succeeded, he'd pay the young man a bonus of $50,000—the rough equivalent of $1,000,000 in today's money, and more than fifty times what was then Tesla's annual salary. Tesla completed the redesign, but Edison rebuffed his attempts to claim payment, positing that it had been an offhand remark and that Tesla had no knack for "the American sense of humor." This enraged Tesla, who quit in a fit of rage and vowed to create his own electrical power system to compete with Edison's DC offering. Unfortunately, in his anger, Tesla had failed to consider his material needs—New York has always been an expensive place—and in order to make ends meet, Tesla soon found himself taking a menial job digging trenches, laying none other than Edison's DC electrical wires.

Tesla was down, but he was not out, and as he labored, he came up with a novel and brilliant idea about how to use a revolving magnetic field to build a high-efficiency, multiphase, AC-based motor. Tesla successfully demonstrated the revolutionary idea and obtained a patent for it, and his invention was quickly noted by Westinghouse, who had just begun to utilize alternating current to compete with Edison in the AC-DC War. Tesla convinced Westinghouse to pay an astonishingly high royalty fee to license Tesla's technology—$2.50 for every horsepower in every AC motor Westinghouse's company sold. But the fee proved worth it, as Tesla's motor became instrumental in swinging the needle toward AC in the War of the Currents. Tesla experienced great personal satisfaction and vindication in knowing that his invention led directly to Edison's downfall—in fact, he drew such pleasure from this that he eventually tore up his licensing contract with Westinghouse, letting the company have his patent for a far more modest sum.

The Chicago Columbian Exposition of 1893 brought Tesla's multiphase AC power grid design to the forefront of electrical power technology. After viewing his AC power system, the whole world came to adopt his technology, and Tesla began to be treated as something of an international celebrity. His extensive knowledge and unique flair for showmanship—coupled with his 6-foot 2-inch height and good looks—powered him to stardom. Tesla gained a reputation in popular culture as the archetypal "mad scientist," which holds even to this age. He was granted over 800 patents over the course of his unique life.

Unfortunately, Tesla fell upon the sword of his own grandiosity. In his late thirties, he grew captivated by Hertz's discovery of the electromagnetic wave and became obsessed with the idea of wirelessly transmitting large amounts of electrical power. He used his fortune from the initial Westinghouse license to work on the idea in his own laboratory in Colorado Springs; he then obtained further investment from Edison's old backer, J. P. Morgan, to construct a huge wireless tower at Wardenclyffe, New York. He boasted that using that facility, he would be able to transmit significant electrical power all the way across the Atlantic Ocean to Paris, without using cables.

Tesla's idea was flawed, and he was never able to fulfill his promise. By 1903, he was broke, and Morgan refused to provide any more funds. Washed up at the age of forty-seven, Tesla spent forty long years in the second half of his life living like a dragonfly skimming over the surface of a great body of water. He became enamored, in succession, with a string of mysterious and unlikely technologies, including the "wireless death ray," "technology for communicating with space aliens," "oceanic thermal energy harvesting," and "gravitational kinetic energy." Meanwhile he ridiculed Edison as a "mere inventor" and Einstein as being a manipulator of mathematics to fool people. Even though there were flashes of brilliance in his free roaming ideas, none of his later efforts ultimately yielded any inventions with significant, direct impact.

Tesla never married, and when he died in 1943, he was friendless and penniless, suffering under an unstable mental state and attended to by only his closest associates in his late life: the pigeons living outside his New York City hotel room window.

Of the three great late nineteenth-century electrical engineers who had huge impacts on society, only one was not well known in America—but the company that bore his name is known throughout the world.

Werner Siemens was born to a large family of farmers in Prussia in 1816, the fourth of fourteen children. His family was poor and couldn't afford to send him to school to get an education, so at age sixteen, he left home and joined the Prussian army, where he received special technical training. After William

Cooke and Samuel Morse introduced the telegraph, the Prussian military came to recognize its importance as a communications tool on the battlefield and decided to set up a team to develop its own telegraph technology. Siemens, with five years of electrical training at this point, was named one of the lead engineers.

When Siemens was thirty-one, he successfully developed a new variation of telegraph known as the pointer telegraph. It was not only more reliable than most other equipment then in use, but also lighter and more compact, key attributes in the eyes of the Prussian army brass. With encouragement from his superior officers, Siemens left the military and started a business with a friend with the pointer telegraph as their core product. Within a year, Siemens' company had obtained a contract from the government to construct a telegraph system from Berlin to Frankfurt.

From the beginning, Siemens' business vision was not limited to Germany alone. Rather, he envisioned his telegraph business stretching across the entire world. He leveraged a great resource: a network of reliable, trustworthy employees—his family—to build his company, assigning his brothers to various international outposts to help develop Siemens' business abroad. Through his brother in Russia, he won the Baltic Sea-to-Black Sea telegraph project. Another brother in Great Britain ultimately obtained for the company the big London-to-Calcutta contract. The Siemens Company produced all its own telegraphic equipment and cables, and it even built a special ship, the *Faraday*, that laid a total of six transatlantic telegraph cables. Thus, the Siemens Company was vertically integrated, able to provide all parts of a telegraph solution from equipment manufacturing to installation.

Werner von Siemens. Mondadori via Getty Images

Although Siemens managed an immense commercial business empire, he remained always an excellent hands-on engineer. He contributed to the design of the first loud speaker, and in 1867 published an insightful paper proposing the use of powerful electromagnets to replace the weak permanent magnets then used in all generators. This advancement helped overcome a major technical hurdle and allowed for the first use of generators to replace batteries in telegraph systems, as well as in the electroplating industry.

Following the success of the progressive innovations of Gramme, Edison, and Westinghouse, the size of the electrical generator and motor market ballooned, and Siemens grew to become the world's biggest electrical equipment supplier. At the same time, his company was growing into the role of global leader in applying electric power to transportation systems. Siemens introduced a series of impressive engineering firsts, including the first street trolley in 1876 and the first electric-powered train in 1879. His firm went on to construct the European continent's first underground railway in Budapest in 1896 and mainland Asia's first street trolley line in Beijing, China, in 1899. Truly, Siemens' influence came to be felt across the entire world. When J. P. Morgan reorganized GE, it was based on Siemens' model. When the Japanese economy began to flourish during the Meiji Restoration, the most successful business elites imitated Siemens in establishing enormous, vertically integrated enterprises, which formed the backbone of Japan's powerful electrical industry.

In addition to his commercial and technological achievements, Siemens had long-lasting influence in the spheres of corporate and societal norms. He instituted pioneering ideas in management, including the notion of employee participation in profit sharing in 1858 and the staff retirement fund and life insurance subsidy system in 1872. No company or government at that time had established such a sophisticated welfare system for workers. Siemens was also the first to establish an 8.5-hour working day and the 50-hour work-week policy for his workers. In recognition of his successes, Werner Siemens was granted an honorific title by the German government and thereafter became known as Werner von Siemens. When he died in 1892 at the age of seventy-six, he left behind a legacy of outstanding engineering, entrepreneurship, and industrialism and cemented his place as one of the giants who brought the blessings of electromagnetic technology to all mankind.

ONE HUNDRED YEARS OF ELECTROMAGNETISM

In the hundred-odd years from Volta's 1800 invention of the battery to Marconi's 1901 demonstration of transatlantic wireless communications, the continuous discovery of new applications for electricity and magnetism led the field to evolve from a mere curiosity to an indispensable aspect of human existence. Volta, Faraday, Maxwell, Hertz, and countless others laid the foundations for the science of electromagnetism. Meanwhile, inventors and entrepreneurs like Morse, Bell, Vail, Marconi, Edison, Westinghouse, Gramme, and Siemens successfully utilized electromagnetic technology to bring convenience and comfort to everyday life, from communications to lighting and from entertainment to transportation. Newly emergent enterprises like Western Union, AT&T, Siemens, GE, Marconi Wireless Telegraphy, and Westinghouse all became pillars of the world's economy, and this three-link chain—science, technology, and business—became the highway to real invention and innovation.

The first annual meeting of the International Electrical Congress, or IEC, was held in Paris in 1881. As a way of honoring the outstanding contributors to the scientific understanding of electromagnetism, the IEC decided to name the standardized unit of electrical potential "volt," after Volta. Similarly, the unit of electrical current came to be known as the "ampere" or "amp" after Ampere, "ohm" became the measure of electrical resistance, and "coulomb" became the unit of electrical charge. In subsequent meetings, it was decided that the unit of electrical capacitance would be called the "farad," electric induction would be known as the "henry," and the standard name for the measure of electromagnetic wave frequency would be the "hertz." In magnetism, too, the basic units became the namesakes of early scientific explorers; Oersted, Gauss, Weber, Tesla, Siemens and Maxwell were all commemorated in kind. This became a special way to pay tribute to the forebears who laid the groundwork for mankind to harness the power of electromagnetism.

By the early days of the twentieth century, the applied science of electromagnetism had become a part of everyday life. When the weather was too hot, a person could turn on the air conditioning, and vegetables could be made to retain their freshness by being put into the refrigerator. A turn of the faucet brought an endless stream of clean water pressurized by the electric pump, to say nothing of the easy flick of the finger that could dispel the darkness of night. A person could turn on the phonograph and in an instant be serenaded, even when alone, and family members could converse with loved ones sitting miles away. Even up-to-the-minute news updates from cities around the world were there for the taking, and all of this was thanks entirely to electromagnetism.

Despite these advances, there remained many basic principles of electromagnetics that were still not understood. For example, what actually is electric current? What is it that flows through electric wire? At the time, nobody knew. Furthermore, a number of other seemingly unconquerable technical problems remained. How, for instance, could the range for long distance telephone calls be extended so that the analog voice signals would not deteriorate beyond recognition? And could voice or music—more than just the blips and beeps of the Morse-coded telegraph—be transmitted using wireless technology?

These were just dreams, but in time, solutions to these important and difficult technological challenges would need to be—and were—found. However, they came from a wholly different area of research. The Age of Electromagnetism ended for humanity with the 1800s. The dawn of the twentieth century ushered mankind into a new era of invention: the Age of Electronics.

II

AGE OF VACUUM ELECTRONICS

6

CURRENT FLOW IN A VACUUM

Though scientists had the ability to anticipate and reliably predict electrical phenomena with ever-growing certainty since the work of Volta at the dawn of the nineteenth century, the nature of electricity—the fundamental units of physical matter that carried electrical charges—remained a mystery for generations. Of course, from the very inception of the study of electrical phenomena—from the days of Gilbert and "De Magnete"—researchers eagerly sought the key to unlock this puzzle.

A few years after Volta invented the battery, Humphry Davy, the predecessor to Michael Faraday at the Royal Institution, came across the first clue. Davy famously constructed an enormous battery of his own in order to perform important experiments in electromagnetism. In 1807, he placed two sharpened carbon rods in close proximity to one another and applied a high voltage across the gap. The resulting electric arc that shot between the two points of the rods not only created a bright, white light, but also showed that electrical current could indeed jump across an air gap at the same time. Intrigued by this phenomenon, Davy continued his research. Over time, he discovered that during the creation of this arcing light, the tips of the carbon rods were readily oxidized and consumed—effectively burned up in the air. To minimize this problem, he devised a means of setting the carbon tips inside a sealed glass tube and then vacuuming out all the air inside. His intention was to remove all the oxygen from the tube and therefore, he hoped, inhibit the rods from burning. As it happened, Davy's experiment failed to produce the expected results, but this was mainly due to the crude vacuum technology available at the time, which allowed air to leak freely into the glass enclosure. His idea had merit, but this field of research could not be meaningfully pursued until the coming of major advances in vacuum technology.

CATHODE RAYS

In 1855, almost fifty years after Davy's initial experiment, German scientist Heinrich Geissler invented a high performance vacuum pump that allowed him and his fellow research scientist, Julius Plücker, to resume the exploration launched by Davy. Geissler's improved vacuum design was similar to a ratchet: gases could be slowly pulled out of a tube, but once drawn out, they couldn't re-enter. Rather than a mechanical cog and pawl, however, Geissler's ingenious design used the diffusion of mercury vapor to displace gas molecules in the enclosed tube.

In addition to being an outstanding experimental physicist, Geissler was also an excellent glassblower, and he was able to perfect the art of sealing metal tubes through glass walls so as not to allow air to leak at the joints. This enabled Geissler and Plücker to conduct many innovative electrical discharge experiments. (Years later, it would also allow Edison and others to build the light bulb—a hot filament sealed in a glass enclosure.)

At about the same time Geissler built his improved vacuum tube, another German engineer named Heinrich Daniel Ruhmkorff introduced an induction-based technique that could generate extremely high voltage to allow for more meaningful and controlled research into the phenomenon of electrical discharge in a vacuum. Utilizing Ruhmkorff's high voltage source alongside his own superb vacuum technology, Geissler was able to break new ground in the study of vacuum electronics. Like Davy, Geissler observed that when the voltage between two carbon tips enclosed in an un-evacuated glass tube was raised to a certain critical level, an arcing, bright light would pass across the gap, accompanied by a simultaneous flow of electrical current between the two tips. But pushing the experiment further, Geissler found that as he used his mercury pump to suck air from the tube, the light emitted by the paired carbon rods would gradually weaken. At low enough air pressure, the light would eventually cease altogether, and the flow of electricity would stop along with it. The discovery was itself interesting, and Geissler decided to press on. He first pulled a high vacuum from the tube, and then backfilled different gases into it. To his surprise, in the presence of these other gases, the flow of electricity resumed, and with it, the arcing light. However, thanks to the presence of these backfilled gases, the color of the new emitted light was different! In the presence of hydrogen gas, the emitted light glowed not white, but red. With a gaseous sodium vapor in the tube, the airborne electrical pathway glowed yellow. No one could have predicted that filling the tube with different gases would induce different colored lighting! Plücker even measured the spectra of various gas emissions and discovered many unique features which could not be explained by Newtonian physics.

Following Geissler's pioneering work, the study of electrical conduction and light-emitting phenomena became a popular area of research for many scientists, particularly in Britain and Germany. The principal question on researchers' minds was: what was responsible for the electrical conduction in the vacuum between the two electrodes? Was it clusters of invisible charged particles shooting across the vacuum? Or was it residual, charged gaseous molecules, known as ions? Was it perhaps some kind of wavelike oscillation between the two electrodes, like electromagnetic waves? Or was it something else altogether?

As vacuum technology advanced, high-voltage power sources and electrode materials improved in lockstep. In 1878, British physicist William Crookes discovered that even though light died away and current ceased to flow under high vacuum conditions, there was a certain voltage threshold beyond which a sustainable flow of electrical current could resume, albeit without inducing glowing light. The magnitude of this resumed current was such that it could not be accounted for by any residual gaseous ions in the tube. Crookes hypothesized that the current was instead carried on a cluster of small, charged, invisible particles that shot out in a straight line from the cathode—the electrode from which the particles originated—toward the other electrode, which was known as the anode. He named these hypothetical, invisible, charged beams "cathode rays." In order

Crookes cathode ray experiment with Maltese Cross (replica). Andrew Lambert Photography/Science Source

to attempt to prove the existence of these cathode rays, he mounted a Maltese cross made of metal plate and mounted inside a glass tube. He then coated the inside glass wall behind the cross with phosphorescent material which would glow under cathode ray impingement. When Crookes powered on the cathode ray, the space at the rear of the tube glowed, except for an area cast in shadow directly behind the cross. This made clear that the invisible particles of the cathode ray had been blocked by the cross, lending great credence to Crookes' concept.

Crookes' work appeared to strongly support the hypothesis that electrical conduction in a vacuum was due to beams of charged particles, and the British scientific community at large began to throw its support behind his theory. However, as Crookes continued his research, he found that while externally applied magnetic fields could deflect cathode rays, for some unknown reason, electrical fields appeared to exert no influence upon them. This seemed to discredit the charged particle theory and lent credence to the notion that conduction was carried not by particles, but by a yet unidentified wave phenomenon. In fact, the German-Hungarian scientist Philipp Lenard even proved experimentally that cathode rays could "tunnel" through thin layers of aluminum film, itself a unique property of a wave that could not be explained by the particle hypothesis. This led the German scientific world at large to come to support the idea that cathode rays must not be streams of charged particles, but a form of waves.

THE ELECTRON EXPOSED

As the debate over cathode rays grew, more researchers entered the field, each generating ever-increasing controversy. Conflicting reports of observed phenomena abounded, and the quantitative truths governing the principles of electrical conduction remained shrouded in mystery. National pride and Anglo-German politics played a significant role in inserting dispute into this scientific search as well.

Finally, in 1894, British physicist Joseph John (J. J.) Thomson decided to get to the bottom of this issue and determine once and for all whether cathode rays, the carriers of electrical current, were particles or waves. At the time, Thomson was the Cavendish Chair Professor at Cambridge University, a position set up by James Clerk Maxwell himself, and with this title, J. J. Thomson was widely acknowledged to be the leading scientific authority in all of Great Britain.

As he sought out a definitive answer, J. J. Thomson sat every day in quiet contemplation in a rocking chair that Maxwell had bequeathed to him. As he rocked, he devised and planned a series of experiments specifically designed to resolve the controversy. In the first set of experiments, Thomson proved

J. J. Thomson. Courtesy of Master and Fellows of Trinity College, Cambridge

indisputably that cathode rays carried a negative charge. In the second set of experiments, using the most advanced vacuum technology at the time, Thomson managed to prove that Crookes and others were wrong in one key respect: electrical fields could indeed deflect cathode rays, and the direction of that deflection was consistent with the notion of cathode rays carrying negative charges. Previous experiments by others, Thomson proved, utilized inadequate vacuums, a condition that allowed residual anomalies to conceal the influence of external electrical fields on the cathode ray.

In his third group of experiments, Thomson meticulously measured the precise magnitude by which cathode rays could be deflected by the application of different levels of magnetic fields. Using those experimental data and Maxwell's electromagnetic theory, he was able to derive the electrical charge-to-mass ratio (e/m) of the fundamental, negatively charged particles that formed the cathode ray itself.

On April 30, 1897, J. J. Thomson took the lectern so often graced by Davy and Faraday in the past and gave a lecture at the Royal Institution. In that lecture, he announced the groundbreaking results of his research: cathode rays were formed by negatively charged particles originating from the cathode, and no matter what methods were used to generate those particles or the nature of the cathode material used, all particles in all cathode rays were identical. It didn't matter if the source was a battery or a generator, or if the cathodes were made of carbon or metal—all particles in all cathode rays were the same. Based on the measured e/m ratio and assuming the negative charge of any individual particle in a cathode ray was equivalent and opposite to that of a positively charged hydrogen ion, an already known quantity at the time, then each cathode ray particle's mass was only 1/1837th that of a hydrogen ion, making cathode ray particles the lightest particles in known existence. This last finding was particularly shocking, because it implied that each particle in a cathode ray was just a tiny fraction of an atom, and up to that time, it was believed that an atom was the smallest, most indivisible particle in the entire universe. Could there really be particles even smaller than an atom? Thomson postulated that there were.

Thomson's discovery and announcement shook the scientific world, and his results were independently verified twelve years later in a brilliant experiment by physicist Robert Millikan that directly measured the precise amount of negative charge on each particle, which indirectly also revealed the mass of the particle. A clear picture at last emerged: the negatively charged, extremely light cathode ray particles were part of an atom, and they constituted the most basic carrier unit of electric current. These electrically charged particles were thereafter named *electrons*. Normally, these electrons were held inside a matrix of solid atoms in the cathode, but when confronted with a strong enough electric field, some of the more loosely held electrons would no longer be able to "hide" inside the atomic solid, and they would be forcefully pulled into the vacuum, totally exposed. That was when their true nature could be revealed under the watchful eyes of a man like J. J. Thomson.

In time, Thomson found that if the cathode materials were heated to elevated temperatures, then their electrons could more easily be pulled free into the vacuum. Furthermore, he discovered that these electrons could escape into a vacuum more easily still and at even lower temperatures when light of a certain wavelength was shone onto the cathode surface. This phenomenon became known as the *photoelectric effect*, and though it was little more than a side note in the matter at hand, the discovery would later prove to have very important applications.

J. J. Thomson's experiments were rightfully hailed as a landmark accomplishment, and he was and is generally regarded as the discoverer of the electron. He also brought the long-standing dispute between the British and German scientists regarding the nature of cathode rays to a temporary truce. In 1906, J. J. Thomson was awarded the Nobel Prize, and when he died in 1940, he was honored and buried in Westminster Abbey alongside the remains of such other scientific luminaries as Isaac Newton and Charles Darwin.

Though it was revolutionary, Thomson's work still left behind many unresolved issues. He was, for instance, unable to answer why electrons were able to "tunnel" through thin aluminum films, which is a property only associated with waves and not with particles. Despite Thomson's Herculean efforts and his impressive results, the dividing issue of whether electrons were particles or waves still could not be put to rest. It would take another twenty years of research in physics to finally resolve the puzzle with the discovery of an ingenious and counterintuitive truth: electrons could simultaneously behave as both particles and waves! This so-called *wave-particle duality* would represent a major conceptual breakthrough in physics, and it would in time become the foundation of a new branch of physics—quantum mechanics, which allowed mankind to understand the behavior of matters at the atomic level.

THE PUZZLE OF PENETRATIVE LIGHT

Two important, practical inventions were born of cathode ray research. The first was the cathode ray tube itself, or the CRT. The other was the X-ray.

Wilhelm Roentgen. Jean-Loup Charmet/Science Source

The CRT was invented by German scientist Karl Ferdinand Braun in 1897, the very same year that J. J. Thomson measured the electron's charge-to-mass ratio. Using a magnetic field to control the individual cathode ray particles much as wooden fences corral and direct cattle in a stockade, Braun developed a technology that could collimate and focus the cathode rays, then scan the beam onto a fluorescent screen placed at the end of a tube. Braun's device mimicked Crookes' experimental setup with the Maltese cross; it differed in that in Braun's CRT, the cathode rays could be effectively focused, directed, and steered by external magnetic fields. At the outset, this device found no applications; only over time did the many uses of the CRT become fully realized.

The discovery of X-rays, on the other hand, was both quite a bit more dramatic—and fortuitous—and immediately more practical. In 1895, German physicist Wilhelm Roentgen was engaged in high-voltage cathode ray research. His contemporary, German-Hungarian physicist Philipp Lenard, had just discovered the ability of cathode rays to tunnel through a thin aluminum film, thus proving, so he claimed, that cathode rays were waves. This discovery gave Germany's scientific community significant confidence in their dispute against the British over the nature of electrons, so Roentgen, as part of his research, planned to repeat and attempt to improve upon Lenard's experiment.

In preparing his experiment, Roentgen assembled the most advanced equipment available at the time, including an extremely high-voltage power source and an anode made of tungsten that was specially fabricated to withstand extremely high temperatures when high-energy electrons impinged on it. During the experiment, he observed that some of the high-energy cathode rays would hit the side walls of the glass tube, creating a fluorescent glow. This, he feared, might obscure the weak optical signal he intended to watch for. Thus, in order to enhance the sensor's optical contrast and make data collection easier, Roentgen turned his whole laboratory into a darkroom. He also shrouded the entire experimental setup with black cloth, leaving just a tiny hole at the rear to allow weak cathode ray signals that tunneled through the aluminum foil to pass through and be detected.

Before beginning his experiment, Roentgen coated a piece of cardboard with cathode ray–sensitive fluorescent material. He cut off a small piece for use in the experiment and left the remainder on a chair in his lab. As he began to conduct his experiment, he noticed from the corner of his eye something glowing in the darkened room: it was the rest of the cathode ray–sensitive cardboard that he had left on his chair! What could be causing that? When he turned off the electrical power to the cathode ray tube, the cardboard stopped glowing. Clearly, some kind of radiating energy was being emitted from the experimental apparatus that penetrated the black cloth shield.

Roentgen was puzzled. He was convinced that something had come out of the apparatus during the experiment and shot through the room. But it didn't seem that the source of this mysterious energy could be the cathode ray itself, because cathode rays had been proven to dissipate within just a few centimeters under atmospheric conditions. Roentgen ran the experiment again, now trying to block the mysterious energy with a thick book placed in front of the coated cardboard, and still, he found that the cardboard glowed. Whatever it was, this mysterious energy was penetrating through even a thick book! At the same time, he noticed that all the film negatives in his laboratory were now exposed—ostensibly by whatever this new radiating energy was.

Roentgen knew he had discovered a new phenomenon. Without telling anyone, he threw himself into solving the puzzle. Day and night, he systematically carried out a series of experiments, until at last, on Christmas Eve, 1895, he was ready. He invited his wife into his laboratory and started up his apparatus. Roentgen placed her hand on a fresh photographic plate and exposed it for fourteen minutes. When he developed the film, he saw a picture of his wife's hand with her ring on her finger. However, this was no ordinary hand. Rather, it showed only bones and a ring that seem to hang suspended around her finger. Mrs. Roentgen's immediate reaction was, "Oh, heavens! I've seen death!"

After Roentgen publicized his discovery, the photograph of his wife's hand became a worldwide sensation. However, when asked about the nature of the radiation, Roentgen freely admitted that he didn't fully understand it. "I don't know quite what it is," he said, "so I am just calling it an 'X'-ray," the letter "X" carrying the connotation for things unknown.

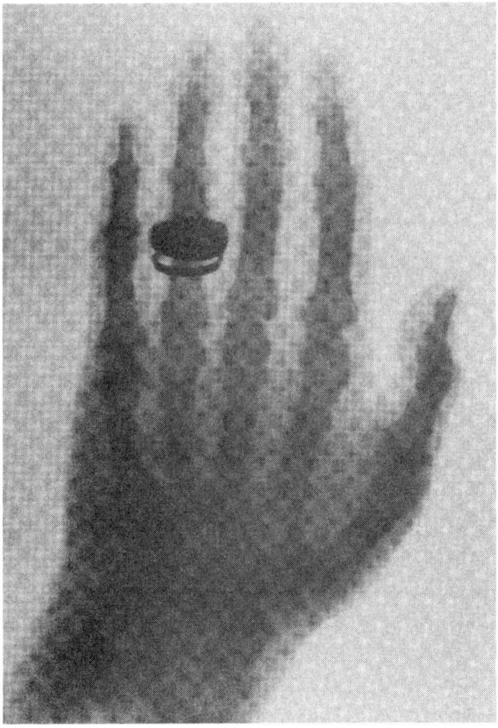

The first public demonstration by Roentgen of X-ray imaging of a hand. Science Source

Despite not comprehending the true nature of his discovery, Roentgen instinctively knew that the technology could be important to the field of medicine. Thus, like the dedicated, idealist scientist that he was, he decided not to apply for a patent, instead offering his work to the public domain. Siemens and GE quickly began to develop practical X-ray equipment in order to help doctors diagnose bone breaks, marking the birth of the electronic medical imaging industry.

Over time, the principles of X-rays came to be understood. In Roentgen's experiment, the high-energy electron beams of the cathode rays collided at high speeds with the tungsten anode. As tungsten atoms in the solid abruptly "braked" the speed of the high-energy electrons, their energy was released in the form of electromagnetic waves with extremely high frequency, or ultra-short wavelength. These were the X-rays. Like visible light and radio waves, X-rays were just another form of electromagnetic waves, and they obeyed all of Maxwell's equations. Quantum mechanics would later show that electromagnetic waves of very short lengths, like X-rays, had very high energy, and they could thus penetrate books, flesh, or even steel. In 1901, Roentgen was awarded history's first Nobel Prize in Physics.

Roentgen's discovery of the X-ray was a classic case of serendipity. Had Roentgen been asked to invent a light source that could penetrate the body, he would have had no clue how to proceed, and the X-ray would not have been discovered for a long time. Luckily, Roentgen readily realized the value of his happy accident and was able to seize upon it. Interestingly, other scientists, including Tesla and Lenard, later claimed to have observed the X-ray phenomenon before Roentgen, but none had ever pursued a deeper course of systematic inquiry into the matter. Thus, Roentgen is rightly credited with the discovery, just as Columbus had been. As Louis Pasteur, the father of microbiology once quipped, *accidents favor the prepared mind!*

THE LEGACY OF VACUUM ELECTRONICS

In the long process of research in vacuum-based electronics, Thomson proved that, just like the water molecule (H_2O) was the fundamental unit of flowing water, the negatively charged, subatomic electron particle was the fundamental carrier of electrical current. But Thomson was not alone in birthing the field of vacuum electronics: Geissler's early work was able to lay the foundation for those who followed. Crookes' work on cathode rays directly led to Braun's CRT, which would serve as the driving technology behind television and computer display monitors until the end of the twentieth century. It would also lead to the creation of many forms of colored lights, such as neon lights, and later to modern day plasma displays. Finally, Roentgen's discovery of the X-ray ushered in the application of electromagnetic technology in medical diagnostics, starting the medical electronics industry.

In time, there would come to be several other important applications of vacuum electronics, such as the electron microscope, which leverages the wave behavior of electrons for ultra-high resolution microscopy. Another important technological innovation would be the use of interactions between electron beams and magnetic fields to generate powerful microwaves. Based on this principle, Albert Hull would in 1921 invent the magnetron, and in 1938 Sigurd and Russell Varian would invent the klystron. These "microwave vacuum tubes" would go on to form the core of radar and satellite communication systems.

In addition to paving the technical foundation for many important applications to come, vacuum electronics also catalyzed a new era in physics. Many of the observations from vacuum experiments could not be explained by Newtonian theories: for example, the emitted light spectrum of an incandescent bulb, the emission spectrum of gas discharge, or the generation of the X-ray. It was obvious that new physics was urgently needed to shed light on all these new observations. In 1900, the German scientist Max Planck proposed his model of blackbody radiation, which precisely predicted the observed light emission spectrum of a hot body, such as a heated filament or the sun. Planck's contemporary, Niels Bohr, would later develop a theoretical model of a hydrogen atom that explained and predicted the observed emission spectra of the hydrogen gas. Planck and Bohr's models introduced two revolutionary concepts into physics: the quantization of energy due to wave-particle duality and the use of probability theory, and their work helped to usher in the era of quantum mechanics, the "new physics" that provided a theoretical understanding of the properties of matters at their most fundamental atomic level. Another important discovery was that the origin of magnetism was directly traced to the movement of electrical charges, including the self-spinning process. So the fundamental origin of all electromagnetic phenomena is related in one way or another to the electron and its motion. All of this new knowledge would eventually lead to the birth of the semiconductor-based, solid-state electronics industry, which enabled the iPhones and iPads of our modern age.

7

CONTROLLING THE FLOW OF ELECTRONS

Although research in vacuum electronics led to numerous theoretical breakthroughs and a host of impressive applications, it did not directly impact the discovery of two of the most important devices ever created for controlling the flow of electrons: the vacuum diode and the vacuum triode. To trace their origins, we have to once more go back to the work of that great innovator from the dawn of the twentieth century, Thomas Edison.

THE EDISON EFFECT

In 1882, while Crookes was hard at work on his Maltese cross experiment in England, far across the ocean, Edison was busy constructing his Pearl Street power station. Even as he designed the generation plant that would power lower Manhattan, he and his team continued to tinker and improve his patented incandescent lamp, with an eye keenly focused on improving its reliability.

The main hurdle confronting the light bulb continued to be the issue of its usable lifetime. But the problem was no longer the filament material itself; thanks to Edison's experimentation, the lifetime of the average carbonized fiber filament already exceeded several hundred hours. The shortfall was that as the filament was heated and began to glow, burnt flecks of tiny black carbon particles would coat the inside wall of the glass bulb, making it increasingly opaque until it blocked out its own light almost entirely. These flecks could not be cleaned from the outside, so their existence posed a difficult technical challenge.

Edison experimented with many possible solutions to this problem, including placing a copper plate inside the light bulb above the filament in the hopes that it would shield the glass from carbon particles. When this failed to yield useful results, Edison began to wonder if perhaps applying an electric field to the copper plate might somehow solve the problem. So Edison designed another light bulb with a copper plate above the filament. He then connected the plate to an external wire. During the experiment, Edison applied different voltages to the copper shield to test its effectiveness. As it happened, applying an electrical field to the copper shield did nothing to boost its ability to block the particles of carbon, but it did have an unexpected result. When the voltage applied to the copper shield was more positive than that of the filament, an electrical current flowed through the vacuum inside the glass between the filament and the copper shield. But when the voltage of the shield plate was more negative than that of the filament, no electrical current would flow. In other words, this system allowed current to flow only in one direction, a phenomenon known as current *rectification.*

Though his finding was indeed curious, to Edison, the overall experiment was a failure. He wasn't interested in discovering any new electrical phenomena; he just wanted longer-lasting light bulbs! It would be fifteen more years before J. J. Thomson discovered and identified the electron, so the physics governing this accidental observation of one-way current flow were not yet understood. And even though Edison did share the findings of his experiment with senior technical staff and advisors throughout his organization, nobody was able to explain the phenomenon or suggest an application for its use. So, consumed with his other work, Edison just filed a patent for his discovery of a "unidirectional current rectifying device," as he did with all his inventions, and he moved on to other projects.

For years, Edison's discovery remained deeply buried among his hundreds of patents. As he never published scholarly papers, the academic world knew very little about the unidirectional current rectification phenomenon, or as some called it later, the "Edison Effect." However, Professor John Ambrose Fleming, a technical advisor at Edison-GE's London branch in the 1880s, was familiar with this particular experiment, and he always kept it in the back of his mind.

Interestingly, the concept of one-way current rectification had been observed before Edison. In 1874, German physicist Karl Ferdinand Braun—the same person who invented the CRT and who would later receive the Nobel Prize in Physics along with Marconi—discovered another source of unidirectional current flow: a brass wire making point contact with a natural mineral known as galena, or lead sulfide. Similar to the Edison Effect, this finding was for a long time just a laboratory curiosity lacking any clear useful applications. However, in 1899, the Indian scientist Jagadish Chandra Bose of the Presidency College in Calcutta, discovered that this rectifying effect could be used to detect the presence of electromagnetic waves, and it could also be much more sensitive than the coherer, which was widely used in that application at the time.

Unlike the Edison Effect that used vacuum tubes, the galena rectifier technology was highly unreliable, and its results were difficult to reproduce. What's more, the physics behind the phenomenon were not yet known—nobody knew quite why it worked. Even after Bose used the galena rectifier to receive an electromagnetic wave signal to remotely detonate explosives positioned in front of the Town Hall in Calcutta, the novel wireless receiving technology remained crude and impractical. Bose, of course, had no idea that a similar—and far more refined—technology had been created in Edison's labs.

In 1900, when Marconi commenced his transatlantic wireless experiment, he hired John Ambrose Fleming, the former Edison-GE employee, as his senior technical advisor to design the ultra-high power wireless transmitting station in Poldhu. At the time, Marconi's receiver design relied on coherer technology, which consisted of thin, rod-like flecks of metal particles suspended in liquid. These metal filaments would align upon exposure to an electric field, and the subsequent "smoothing" of the path of the circuit led to a drop in its electrical resistance. Marconi initially used this change in resistance to detect the arrival of wireless electromagnetic signals. However, the coherer suffered from low sensitivity and slow response time due to its mechanical nature, not to mention difficulty in resetting—at outset, the only way to re-randomize the metal bits was to tap the device manually with a hammer! It was easily the weakest link in Marconi's wireless telegraphy system, and when Fleming was hired on, he began to wonder whether there might be a better way to detect wireless signals.

Fleming was intrigued by Bose's discovery of using the rectifying property of galena devices for wireless signal detection. However, he also knew that the galena current rectifier was too unreliable for use in Marconi wireless telegraphy systems. Marconi needed another, more reliable current rectifying technology. A flash came across Fleming's mind—the Edison Effect!—and Fleming was among only a handful of people on earth who knew about it! Fleming was able to seize upon this historical opportunity, linking the Edison Effect with Bose's rectifying detector concept to come up with a major new breakthrough in wireless communication.

THE VACUUM DIODE

In the twenty-some years since the first observation of the Edison Effect, great advances were made in vacuum electronics. Thanks to J. J. Thomson, the fundamental physics behind the Edison Effect could now be fully explained: when the filament in Edison's light bulb was heated to a high temperature by the electrical current, a continuous stream of electrons would escape from its surface into the vacuum glass enclosure, just like boiling water evaporates into steam. When the copper shield had a more positive electrical potential, these free, negatively charged electrons became attracted to it, causing electrical current to flow continuously across the vacuum. However, when the voltage on the shield was made more negative relative to the filament, it would push the electrons

John Ambrose Fleming. © Science Museum/Science & Society Picture Library

away, so there would be no current flow. In 1904, twenty-two years after Edison patented his "unidirectional current rectifying device"—and five years after Edison's seventeen-year patent had expired—Fleming designed and introduced his own current rectifying device: the Fleming Valve, a refined version of Edison's light bulb. Because of its construction—two electrical terminals sealed within a vacuum—the Fleming Valve also became known as a vacuum diode.

The performance of the vacuum diode was robust and reproducible, a far cry from the erratic point contact galena rectifiers. The device found immediate application in Marconi's wireless receiver systems, and it enhanced system sensitivity immensely. The vacuum diode also found wide use in converting AC power to DC power in many power supply designs. The diode was the first practical device capable of controlling and managing the flow of current at the electron level, thus launching the new *electronic age*. It proved to be so important that it brought both fame and wealth to Fleming, who was knighted by the British crown for his scientific contributions. Of course, Fleming's main contribution was to link a piece of obscure work by Edison with Bose's discovery of using Braun's point contact rectifier for wireless signal detection. This "association" process, or "connecting the dots" action, may on its face seem ancillary or supplemental to the work of fundamental research, but time and again, it has been proven to be an important and invaluable element in the history of innovation.

THE MAGICAL THIRD ELECTRODE

The vacuum diode was the harbinger that ushered humanity into the electronic age, but a far more important invention lay just ahead. This next device would truly revolutionize mankind's ability to control and manipulate the flow of electrons, setting the foundation of modern electronics as we all know it today. Interestingly, this revolutionary invention was also to be created rather fortuitously, and it was developed by a young, brash tinkerer named Lee De Forest.

Lee De Forest was born in 1873, the son of a preacher. The whole family moved from Iowa to Alabama when De Forest's father took a job in administration at Talladega College. With the value of education imbued in him from a young age, De Forest grew up an inquisitive tinkerer, and after years of study, he received his PhD from Yale University, where he wrote his thesis on electromagnetic waves.

Lee De Forest was a highly entrepreneurial young man. Only a few years after his graduation, he decided to start a wireless telegraph company to compete with Marconi. Though courageous and determined, De Forest was neither savvy nor shrewd. He was a poor decision maker, had little aptitude for forward

Lee De Forest with his audion. Science Source

thinking, and the manner in which he handled business dealings was problematic and shabby. He was even charged with commercial fraud and almost spent time in jail, though he was ultimately acquitted.

In 1904, after John Ambrose Fleming introduced the vacuum diode, De Forest began to tinker with the technology, vaguely hoping that he might in some way be able to improve its performance. In 1906, he asked a glassblower to insert a third electrode between the cathode and the anode of the vacuum diode. During testing of his three-terminal vacuum tube, he noticed a curious effect: he could influence the magnitude of the current flowing between the cathode and anode by applying voltage to the third electrode in the middle.

Adding a third electrode to a vacuum diode, it turned out, was like adding a tap and a spigot to a pipe. By changing the voltage level applied to the third electrode, De Forest could alter the flow of electrical current; much like twisting a knob can hasten or slow the flow of water from a faucet. Intrigued with what he'd stumbled onto, De Forest did more thinking and then asked the glassblower to make another three-terminal tube, this time bending the wire of the third electrode into the shape of a zigzag grid. His intuition was that the grid pattern might effectively increase the influence of the third electrode by allowing it to cover a larger cross-section of the vacuum space between the anode and the cathode without physically blocking electron flow. As expected, De Forest found that applying voltage to this "grid" electrode produced a much larger effect upon electrical current.

This vacuum triode, or more simply, the triode, has two critically important functions: *signal amplification* and *switching*, both of which have crucial applications for both analog and digital signals. When a small voltage signal is applied to the grid electrode, an amplified replica of the signal is generated as an electric current flowing between the cathode and the anode. This amplification phenomenon provides an ideal solution to the difficult problem of transmitting weak analog signals, such as voice calls, over long distances: by running the signal through a triode, it can be amplified with minimal added noise or distortion.

In addition to its signal amplification capabilities, the triode can also be used as a near-instantaneous on/off electronic switch. By applying a negative voltage to the grid electrode, the current flow between the anode and cathode can be turned off instantaneously. This fast electronic switching function is crucial in the generation and routing of digital signals.

Though it was his invention, De Forest himself did not understand the physical principles that controlled the operation of the triode. He mistakenly thought that the physical principle affecting current flow was due to ionized residual gas in the vacuum tube. In fact, he even named his triode the "Audion" because of this misconception: the "-ion" at the end of the word referred to ions. In addition, he was unable to explain why he had thought of adding a third electrode inside the diode at all, something not even one of the many other brilliant scientists then involved in vacuum electronics research had ever considered doing. He had simply done it.

The triode was a marvel. However, as De Forest neither understood the fundamental principles governing the triode nor had any clear idea of how to use it, this invention languished in limbo for years after its creation, until at last, opportunity came calling in an unusual guise: the telephone company.

VOICES ACROSS THE CONTINENT

The Bell Telephone Company was established in 1877 by Gardiner Greene Hubbard, who recruited the then-thirty-two-year-old entrepreneur and management genius, Ted Vail, as the company's general manager. Though Vail guided the company to great success, he was eventually squeezed out of his leadership position in 1889 by a block of shareholders who were interested almost solely in short-term profit. Following Vail's departure, the quality of service provided by AT&T capsized, as did its market position. This was the inevitable outcome of the combined consequences of poor management and the tens of thousands of independent telephone companies that shot into existence soon after Bell's original telephone patent expired in 1894.

Collectively, these nimble competitors proved a formidable opponent to AT&T. They chipped away at its core business, leading to a sharp drop in the company's market share and profitability. In addition, AT&T management found itself firmly in the crosshairs of the federal government, as Teddy Roosevelt's trust-busting administration threatened to nationalize AT&T or dissolve it outright.

At this critical juncture, a consortium of New York financiers led by J. P. Morgan took back control of AT&T's board. Morgan's first act in power was to personally invite Vail to return and run the company. Thus, in 1907, after an absence of eighteen years, Vail returned to AT&T and immediately revamped its long-term strategy (much as Steve Jobs would do many years later at Apple).

Upon his return, Vail clearly defined two major goals for the company. The first was to convince Congress that in order to guarantee universally high-quality telephone service throughout the United States, the whole country needed to be linked by a unified network standard, and thus, allowing AT&T to operate as a *benign monopoly* was in the best interest of the nation. His second key goal was to strengthen and leverage AT&T's technological advantages to improve the quality of its customer service, thereby allowing it to fend off the challenges of the independent companies and bolster its position.

In pursuit of the first goal, Vail offered sincere and full cooperation with the government, pledging absolute compliance with all regulations and opening up the company to regular inspection by federal auditors. In pursuit of the second, Vail committed AT&T to building up a technical capacity in communications technology that was second to none. This had the added benefit of ensuring that AT&T would never be held hostage to other companies' patents and know-how. Before Vail's return, AT&T was forced to pay a huge sum of money for the rights to use the automatic crossbar switchboard technology developed by a competitor. In 1900, AT&T again had to bury its pride and spend what was then the astronomical sum of $500,000 to buy Columbia University professor Mihajlo Idvorski Pupin's "loading coil" technology for use in its long distance system. Vail was determined that with AT&T's enormous technical capabilities and resources, such marketplace hostage scenarios should never again recur.

One of Vail's key business strategies against the independent phone companies was to leverage its financial and technology advantages to build a large-scale, coast-to-coast, long distance telephone network. Then, rather than lock the other firms out, he would instead let the independent local phone companies lease services from AT&T under the terms and conditions AT&T set forth. Over time, Vail foresaw that each company would gradually lose its independence and come to rely entirely on AT&T. With the use of Pupin's loading coil, AT&T was able to extend the reach of its long distance telephone lines from New York City to Denver. Though their goal was to stretch farther west, they were unable to continue to expand because the loading coil technology, which was based on induction principles first developed by Faraday, had reached its performance limit. California and the far west, however, were still rapidly booming, and thus, solving the problem of creating a transcontinental telephone line became AT&T's top technical priority.

In order to strengthen AT&T's technical ability and to ensure its position at the forefront of telephony—and, of course, to work toward solving the long distance telephone problem—Vail expanded the role of the technology department of AT&T's manufacturing subsidiary, Western Electric. He gave the group a clear mandate to produce and innovate, which would eventually lead to the creation of one of the most successful R&D centers of all time: Bell Laboratories. While searching for an answer, engineers from Western Electric's technical department took note of De Forest's vacuum triode and recognized its potential as a high-performance signal amplifier, one that could eliminate the bottleneck choking their cross-continental telephone system. De Forest still had little idea of the full potential for what he'd created, so in late 1911, with his finances in disorder, he agreed to sell his triode patent to AT&T for the piddling price of $50,000. After acquiring the rights, AT&T assembled a dedicated and talented technical team, and within two years, they were able to improve the vacuum triode technology until it reached a level suitable for large-scale, practical use.

Using the triode, AT&T engineers designed a high-fidelity, low-noise amplifier for their voice signal repeaters, with performance far superior to the loading coil. In 1914, AT&T was able to leverage this technology to complete the first transcontinental telephone line from New York City to San Francisco. At the inauguration ceremony, Vail invited the aged Alexander Bell and his old partner Watson to speak to each other through the wire. With Bell in New York and Watson in San Francisco, the two men spoke to each other for the first time in over thirty years—only this time, they stood not just a few yards away from one another; they were separated by an entire continent.

8

RADIO

CHRISTMAS EVE, 1904

Thanks to AT&T's investment, vacuum triode technology rapidly matured, and very soon, a wide array of triode products became commercially available. Many engineers began to experiment with the amplification and switching properties of the triodes in different circuit designs. These activities ushered in countless new possibilities, including the long-held dream of sending and receiving voice and music not only through wires, but also through free space via electromagnetic waves.

Though few knew it, music and voice had indeed been successfully broadcast before the adoption of the triode by a man named Reginald Fessenden. Fessenden was a Canadian who, like most other great contributors to the world of electromagnetic technology, had been enamored of electrical currents and electromagnetic waves from the years of his youth. He never formally graduated from a university, but his knowledge and insight about electrical technology were unmatched.

Fessenden joined Edison's laboratory in 1886 as a junior technician. As an extremely creative and competent engineer, he deeply impressed Edison. In addition to his assigned tasks, Fessenden was obsessed with the idea of transmitting the human voice and music using electromagnetic waves. Edison did not support this idea, and he even openly stated that such a feat was technologically impossible. Thus, sensing that he'd never have a chance at Edison to pursue his dream, Fessenden left the company. He worked at Westinghouse, improving the design of AC generators, and he later taught as a professor at two different universities. For several years, he even joined the United States Weather Bureau

Reginald Fessenden. © Bettmann/CORBIS

to head a department dealing with the express delivery of weather information. However, no matter where he was, Fessenden's mind forever dwelled on the challenge of transmitting voices and music wirelessly.

By 1903, through years of hard work, Fessenden was able to gradually develop a complete system concept of how voice and music signals might be sent via electromagnetic waves over a broad area. He envisioned encoding the undulating audio signals, not via an electrical current as with the telephone, but on continuous, high frequency electromagnetic waves appropriately named "carrier waves." These waves, carrying the encoded signals within them, could be broadcast from one point and then received by multiple receivers far away. (See sidebar on AM and FM principles on page 149.) Though Fessenden's breakthrough system concept was correct, by that time, neither Fleming's vacuum diode nor De Forest's triode had yet been invented. People were still using Marconi's "spark gap" technology to generate electromagnetic waves, and neither their frequency purity nor their stability could meet the minimum requirements of the carrier waves Fessenden envisioned.

Fessenden realized that if he wanted to prove his radio broadcasting system concept, he would need to invent his own method to generate carrier waves. He came up with the idea of using a super-high-speed alternating current generator—

an alternator—as the source for his electromagnetic carrier waves. This approach met most of Fessenden's needs: the electromagnetic waves it generated were of very high power, were continuous, and maintained a pure and constant frequency. The primary weakness of the alternator was that the frequency of the waves it generated was quite low. This was a natural limitation due to the mechanical nature of the alternator's rotating armature. Still, Fessenden persevered. He ordered from GE a specially designed, super-high-speed alternator that could produce a record 50 kW of continuous electromagnetic waves at a frequency of 90 kHz. Even though this frequency was much below what Fessenden had originally conceived of, it still served well enough to perform a conceptual demonstration.

Another major technical obstacle was the receiver. Coherers were desperately unresponsive, as was already known, and neither a reliable galena point-contact rectifier nor the vacuum diode were yet available. To solve the problem, Fessenden invented a thermal device that he called a barretter. The barretter functioned by detecting small fluctuations in electrical resistance as the temperature of the resistor changed slightly as the energy from wireless electrical signals was absorbed. Even though the barretter's performance was only marginally adequate, it was an ingenious solution to this difficult problem.

By this time, Fessenden had secured investments from a financial group in Pittsburgh to establish a wireless telegraphy company that aimed to compete with Marconi. Though he did manage the routine wireless telegraph business as promised, Fessenden devoted the bulk of his energy and time to planning what he hoped would be the world's first successful radio broadcast. He built a complete broadcasting system, using his powerful alternator to create carrier waves and encoding the sound signals via amplitude modulation with a modified carbon granule microphone. Finally, using the barretter, he designed and built several receivers and installed them on Navy vessels and fruit transport ships traveling to and from Central America.

On Christmas Eve, 1904, Fessenden made history's first radio broadcast from the little burg of Brant Rock, Massachusetts. In the beginning, speaking into the carbon microphone that modulated the carrier wave, he wished everyone a Merry Christmas and then played Handel's "Largo" from a phonograph. He couldn't be certain whether or not his broadcast was being heard, but he continued all the same. After broadcasting an evening's worth of programming, Fessenden himself played the Christmas carol "Oh Holy Night" on his violin. Though he didn't yet know it, his efforts were not in vain: Fessenden's broadcast was received as far as 200 miles away by ships off the Virginia coast. Fessenden's long-held dream was finally realized: human voice and music were indeed sent and received using wireless technology!

Even though Fessenden was an exceptionally innovative technologist, he was not a good businessman, and he knew very little about public relations or self-promotion. Although Fessenden accomplished something that even Edison had believed was impossible, this historic event went almost completely unnoticed by the general public. Although Fessenden was indeed a genius, he suffered from one unfortunate shortcoming beyond his own control: timing! He appeared on the scene too early. His system—too bulky, too expensive, and of too low quality—was doomed from the outset, as the truly practical carrier wave generation technology needed to power workable radio broadcasting would remain a dream for a few years to come. Nevertheless, he sketched out the entire radio broadcasting concept and demonstrated its fundamental feasibility.

CORE RADIO TECHNOLOGY

Though Fessenden came along too early to truly usher in the age of radio, another man—Edwin H. Armstrong—was right on time. Armstrong was born in New York City in 1890, just after Edison had grandly brought electrification and illumination there. Hearing stories about what life was like "before," he became fascinated by electromagnetic technology, and while still a youth, Armstrong even began to reproduce some of Faraday's famous experiments.

At the age of seventeen, Armstrong enrolled at Columbia University, where he studied electrical engineering under Professor Mihajlo Pupin, the inventor of the loading coil technology purchased by AT&T to enable early long distance telephone systems. At Columbia, Armstrong remained as enamored of and curious about electromagnetic phenomena as he'd ever been, and during his sophomore year, he began to experiment with using the new vacuum triode in electrical circuits.

During one of his experiments, Armstrong discovered that if he connected a portion of the amplified output signal from the circuit back to the input of the triode amplifier, a process known as "feedback," it would cause regenerative amplification. Unchecked, this created an unpleasant, squeaking noise, but if carefully controlled, the regenerative amplification could have positive effects. Under certain conditions, input electrical signals could be further amplified to produce high fidelity output of heightened quality. Experimenting further, Armstrong found that under still different feedback conditions, the output current could be made to exhibit various stable and predictable oscillations. Over time, Armstrong came to realize that most such circuit behavior could be predicted using the mathematical methods Gustav Kirchhoff had developed for use during the construction of the second transatlantic telegraph cable.

Like nearly everyone in the field, Armstrong knew about Fessenden's experiments with the radio and all the technical bottlenecks that existed at the time. Foremost among the challenges was the effective generation of a high powered, high frequency carrier wave to transmit audio signals. Armstrong, through his own experiments, came to realize that the stable oscillations that triode-based feedback circuits could produce were exactly the kind of high frequency, continuous carrier waves that Fessenden had dreamed of. Realizing the important implications, Armstrong took to perfecting the performance of these feedback circuits, which came to be known as *oscillators*. These oscillators—compact and capable of being tuned to specific frequencies—were ideal sources of carrier waves.

Edwin H. Armstrong as a student in 1912. Courtesy of Edwin H. Armstrong Papers, Rare Book and Manuscript Library, Columbia University Libraries

Previously constrained by the mechanical limitations of the revolving armature of an alternator, the frequency of electromagnetic waves produced by triode-based oscillator circuits could be made very high indeed, reaching upwards of tens of millions of cycles per second (or megahertz, MHz). At such frequencies, carrier waves could indeed carry multiple channels of high quality voice and music broadcasts. In 1914, Armstrong filed a patent on his oscillator invention, marking the beginning of the commercial radio era.

With the technical bottleneck of carrier wave generation removed, GE, AT&T, and Westinghouse all began to develop their own brands of radios, despite the fact that the product was still seen as a novelty with limited market

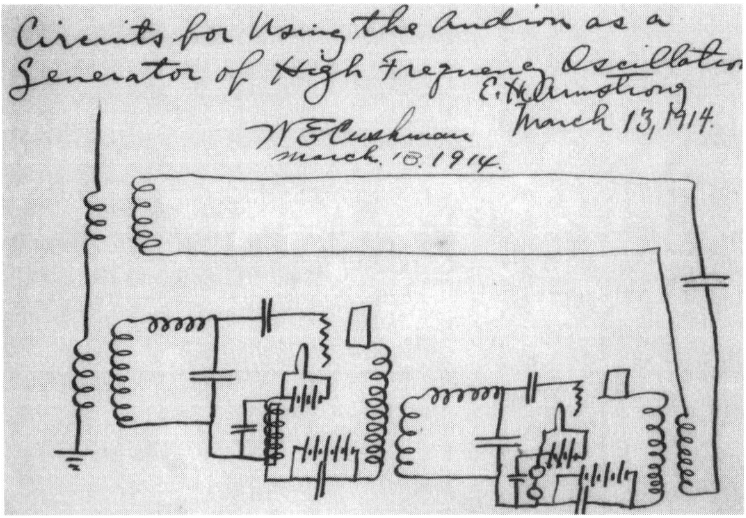

Armstrong's sketch of the original oscillator circuit in 1914. Courtesy of Edwin H. Armstrong Papers, Rare Book and Manuscript Library, Columbia University Libraries

size. Indeed, many amateur radio hobbyists, known as "ham radio" operators, chose to design and build their own radio equipment, using triode oscillators in the transmitters and vacuum diode or point-contact rectifiers in the receivers. Homemade antennae soon sprung up on rooftops everywhere. These hobbyists transmitted and received conversations with each other through radio waves, exchanging information as a faceless, diverse, yet emotionally close-knit social group, just like early adopters of the Internet would be many years later. Of course, there could be no such thing as a private conversation when radio waves were free for the listening. However, one positive outcome of the swelling number of hobbyists was the steady cultivation of engineering talent in radio communications, a skill that would become very necessary within just a few short years.

Shortly after the First World War broke out in 1914, the radio became an essential communication tool for the military and the intelligence community. Suddenly, this cottage industry had become of great significance. When the United States entered the war in 1917, public radio use was banned across the country as a security measure. Thereafter, many wireless hobbyists were drafted and became radio operators in the military. Armstrong himself joined the U.S. Army and was sent to Great Britain to develop ultra-high-sensitivity wireless receivers. The purpose of the project was to detect the weak electromagnetic waves emitted by spark plug firings in engines mounted on distant German airplanes. Even though the project itself was unsuccessful, Armstrong drew much inspiration from it and eventually used it as a springboard to create a critical core radio receiver technology known as "superheterodyne detection."

The basic concept of superheterodyne detection was to create a method by which the frequency of a radio receiver could be adjusted on the fly. If successful, a radio receiver could be made to detect a whole range of radio signals without compromising its performance or increasing its design complexity. Remarkably, the superheterodyne concept was originally conceived by Fessenden, but he was not able to implement it due to the crude component technology he worked with. Thus Armstrong was able to be the first to implement the idea, and in 1918 he received a patent on it. Just like his mentor Pupin had done with his invention of the loading coil, Armstrong sold his patent rights, this time to Westinghouse, making himself a fortune.

RCA AND SARNOFF

Urged on by the needs of the military during the First World War, radio technology developed rapidly in the late 1910s. In addition to radios used by their ground forces, several countries also developed compact, lightweight radios for airplanes so that pilots could communicate with land units and other aircraft in order to coordinate their tactics. The German military also used long distance radio to communicate with their Zeppelin flying ships.

Among all the armed forces, the navy had the greatest need for long distance radio communications because their fleets and men were the most geographically dispersed. The U.S. Navy considered radio technology to be critical to both its mission and the security of the entire nation. By the conclusion of the First World War, the Navy had declared its intent to control the wireless communication industry. Though the American Marconi Company had been operating for some time, the Navy felt uneasy that Guglielmo Marconi, the founder and largest shareholder of the country's most dominant wireless enterprise, was not only a foreign citizen but also a freshly minted Fascist Party member. His radical politics, they argued, put him directly at odds with the needs of America. Furthermore, the Navy also worried that if the limited available frequency spectrum for radio communication was not brought under regulation, the military's access to its use might be jeopardized in time of need. With these arguments, the Navy lobbied the U.S. government for total control of the wireless industry.

After due deliberation, Congress found the Navy's intentions to be too extreme and did not accept them. By 1920, business-friendly Warren G. Harding sat in the White House, and the order of the day was to support private enterprise. Instead, the government encouraged GE to organize a new company named the Radio Corporation of America, or RCA, to specialize in the business and development of radio technologies. In short order, and under the direction and influence of the federal government, RCA acquired the American Marconi Company. Under additional government pressure, even industry giants AT&T and Westinghouse transferred their wireless technology patents to RCA in exchange for a share in the new company's equity. In this way, a near-monopolistic U.S. radio company was created. To alleviate their original concerns, the U.S. Navy was also granted some clearly defined authority in the operation of the new company, specifically for matters in which national security issues were involved.

As RCA was being established, a law was passed stipulating that every individual or enterprise operating a wireless transmitter-receiver must be registered with the government. Companies transmitting wireless signals for commercial use had to apply for a license. Strict limits were imposed on the allocation and usage of frequency ranges to avoid radio interference. As a result of these guideline-defining actions and the rapid maturation of radio technology, more and more private enterprises felt comfortable investing in the radio, and from 1920 onward, the radio and broadcasting businesses took off.

David Sarnoff. Courtesy of Hagley Museum and Library

In 1920, Westinghouse received the first commercial radio license. Its station, KDKA in Pittsburgh, marked history's first private sector broadcaster. Other stations soon followed, most of which filled their daily programming with popular music and real time play-by-play coverage of professional baseball games. Later, stand-up comedy routines, radio plays, and news were added. All of these programs grew to become very popular, and the general public quickly became fascinated by this new mass medium. Thanks to commercial radio's initial success, new stations mushroomed everywhere, and within two years after the inception of the first one, 536 independent radio broadcasting stations were registered to operate in the United States. In these two years, the total number of regular listeners for radio broadcasts exceeded one million people nationwide.

As is often the case with new commercial technologies, during this key developmental period of radio, one individual came to play a pivotal role: David Sarnoff.

David Sarnoff was a Jewish Russian émigré who moved with his family to New York City at the age of nine. While still a boy, he hawked newspapers to help the family get by. Sarnoff was diligent, aggressive, and street-smart, and at fifteen, he found employment as an office boy at American Marconi Company in New York City. At night, he studied wireless telegraph technology in the hopes of becoming a telegraph operator.

In addition to his drive and hard work ethic, Sarnoff was blessed with outstanding interpersonal skills. Every time Guglielmo Marconi came from Europe to the United States for business, Sarnoff would jump at the opportunity to attentively and professionally assist the company chieftain by delivering his messages and taking care of his personal matters. Sarnoff's competence and discretion made a strong impression on Marconi, and they paid off: by the young age of seventeen, Sarnoff was certified to serve as a wireless telegraph operator in the American Marconi Company's New York office.

Sarnoff was a prodigy at self-promotion and catching the eye of those in power. When the *Titanic* sank in 1912, Sarnoff worked tirelessly, transmitting, receiving, and relaying telegrams late into the night. Dealing with the news reporters who covered the catastrophe, he cast himself in a heroic role. He claimed to have stayed at his post nonstop for three straight days, doing everything possible to save the lives of the shipwrecked. Curiously, whether or not Sarnoff was even on duty the night the ship struck that fatal iceberg has never been officially verified. Nonetheless, Sarnoff gained some reputation for himself in the process, and he developed a deep appreciation for the power of the media. This impression would go on to influence his entire career.

Sarnoff was talented and aggressive, and quickly he moved up in the management ranks. When GE set up RCA at the direction of the U.S. government and acquired American Marconi Company, Sarnoff became, naturally, one of the founding employees. His experience and expertise in the wireless business, coupled with his ability to inspire others' confidence in his abilities, quickly earned him the trust and appreciation of the managers sent from GE. Thus, it was of little surprise when in 1921, Sarnoff was appointed to the position of executive general manager of RCA, making him, essentially, the most influential person in the radio industry. He was only thirty years old at the time.

During the 1920s, Sarnoff published a series of articles to articulate his vision for the development of the radio industry. He correctly recognized that in addition to the radio as a hardware product, an even more important dimension would be the program content of radio broadcasts. He prophesied that radio would surpass newspapers to become the most important form of media in the decades ahead, and advertising would be an important source of revenue for the radio business. This level of strategic thinking and insight far exceeded that of his peers, and Sarnoff's position grew even more potent.

After Sarnoff took over the reins at RCA, he made a major investment in developing the most advanced family of radios, which he branded "Radiola." The simplest model of Radiola sold for $75, equivalent to about $800 in today's inflation-adjusted dollars. This was a steep price, but Sarnoff did his best to make the offer attractive. To grow the radio business, RCA sponsored and produced a diverse collection of broadcast programs to attract listeners of a wide variety of interests and tastes. The most sensational of these was the ring-side, live coverage of a world heavyweight championship boxing match between Jack Dempsey and Georges Carpentier, the first ever sporting event to gross over one million dollars in revenue. This boost in programming led to a boom in radio sales, and by 1924, RCA's revenues from the Radiola alone reached $83 million. Soon, it became commonplace for families across the country to sit together after dinner around their radios in their living rooms, listening to their favorite programs.

RCA followed in the footsteps of GE, AT&T, and Westinghouse to become a giant in the electronics industry. In order to expand and control the emerging radio broadcasting landscape, Sarnoff engineered the acquisitions of numerous radio stations from both independent owners and from AT&T. Given RCA's growing dominance in programming, most independent radio stations faced extreme difficulty in competing with the network. Many of them either failed or were acquired by RCA. Sarnoff organized his stations into an interrelated "network" to leverage programming costs-of-scale and increase advertising rates. This same model would later serve as the network-affiliate model put in place during the early days of television.

Sarnoff's acquisition strategy was very successful—too successful, in fact. Within a few years, the RCA network had simply become too large and too powerful, and it began to draw strong warnings from antitrust law enforcers. To avoid sanctions, Sarnoff decided to solve this matter by splitting his behemoth radio network into two systems: the "Red" and the "Blue." The Red network was spun off from RCA as an independent company. It would later evolve into the American Broadcasting Company, or ABC. The Blue network remained an inhouse entity, and it became the National Broadcasting Company, or NBC.

Despite Sarnoff's maneuvering, his company still came under great pressure from the antitrust lobby, and in 1932, the U.S. government decided to take steps to reduce RCA's monopoly in the radio hardware business. The Justice Department ordered the severance of RCA's cross-ownership equity arrangement with GE, AT&T, and Westinghouse, making RCA totally independent. The government also demanded that RCA share its various patents with other companies in order to stimulate market competitiveness. These steps further spurred the growth of the radio industry. By 1933, the number of radio owners in the United States reached thirteen million. Famous radio announcers became household names and quickly achieved celebrity status similar to Hollywood movie stars.

ARMSTRONG'S TRAGEDY

While Sarnoff was reaping the marketplace success of the radio, Armstrong was left with bitter fruit. Armstrong had created a feedback circuit that produced stable, high-power, high-frequency carrier waves, thus removing the major technical bottleneck to sending audio signals via electromagnetic waves. However, when Lee De Forest, inventor of the vacuum triode, learned of young Armstrong's success, he applied for a nearly identical patent, claiming that he observed the oscillation before Armstrong did. When radio sales began to soar in the 1920s, De Forest decided to sue Armstrong for patent infringement. This case was fought in court for almost fourteen years, reaching all the way to the U.S. Supreme Court. In the end, Armstrong lost—but not necessarily because he was in the wrong. Indeed, the technical community rallied behind Armstrong throughout the deliberation, going so far as to award him the Edison Medal of Honor for his invention of the oscillator circuit at the

1934 Electrical Engineers Association meeting, which took place after the case ended. Why, then, did Armstrong lose the Supreme Court case? The answer was simple: money and power. In 1911, De Forest sold his vacuum triode patent to AT&T. Thus, as the owner of the patent—and partner with RCA—AT&T had a vested interest in ensuring that the value of its property was as great as possible. Though Armstrong was armed with the truth, he was fighting a losing battle in going up against this powerful corporation, and his stubborn adherence to principle kept him from accepting numerous opportunities to compromise and settle.

Even though Armstrong lost his oscillator patent case, he kept his innovative spirit. He also maintained all his rights to regenerative amplification technology, for which both Westinghouse and RCA paid him a substantial royalty. This windfall gave him the financial security to propose to and marry his long-time love, Marion McInnis, who also happened to be David Sarnoff's executive secretary at RCA. For her, Armstrong designed and built the world's first portable radio, which he famously unveiled during a day of romantic picnicking at the beach in 1922.

Armed with the love of a good woman, Armstrong's creativity in radio engineering continued to flourish. In 1934, just after the completion of his court case with AT&T, Armstrong invented Frequency Modulation, or FM, a new signal encoding technique that significantly improved radio sound quality. All previous radio transmissions were encoded via Amplitude Modulation, or AM, in which signals were embedded on the carrier waves as a variation in the amplitude of the waves themselves. In FM transmission, audio signals were encoded via modulation of carrier wave frequency, which were far less vulnerable to distortions caused by atmospheric conditions. Though AM technology was already widely in use, Armstrong was certain that FM radio was superior and would be the "wave" of the future.

Armstrong offered to sell the FM patent rights to RCA; Sarnoff, however, was noncommittal. The AM radio represented an extremely profitable product line at RCA, and Sarnoff posited that the new FM technology might undercut its already extant market share and damage the AM business. So Sarnoff dithered, stringing Armstrong along as RCA continued to reap profits in the AM market. Eventually, Armstrong grew weary of waiting for a decision from Sarnoff, and he decided to venture into the product business himself, investing all of his personal fortune in developing and mass producing a family of brand-new FM radios.

This did not go well.

Filled with deep resolve, Armstrong was looking forward to an AM-FM battle between himself and RCA, just like the AC-DC War held two generations earlier between Edison and Westinghouse. Armed with superior technology, Armstrong felt sure that he would win at the end. He threw significant resources into manufacturing and selling his own brand of FM radios. Unfortunately, Sarnoff had foreseen this response, and he had an ace up his sleeve.

Unbeknownst to Armstrong, Sarnoff had quietly begun his own efforts to develop FM technology at RCA. At the same time, he exerted influence behind the scenes with the newly formed U.S. Federal Communications Commission, or FCC, which had the authority to regulate all phases of communications operations in the United States, including technical matters such as radio frequency allocation and licensing. Flexing his political muscle, Sarnoff convinced the FCC to allocate the frequency bands originally set aside for FM to a new, emerging product that RCA was developing at the time, while pushing the frequencies designated for FM broadcasting to a much higher range of the electromagnetic spectrum. Though this represented no great change to the difficulty of FM radio production or radio broadcast quality, it dealt a fatal blow to Armstrong's nascent company, just as Sarnoff had intended. Armstrong had already poured all of his money into manufacturing FM radios based on what he understood to be the FCC's designated broadcasting spectrum. With this sudden change, whole swaths of his finished product had become obsolete. Just like that, Armstrong's life savings went up in smoke.

No stranger to the courts, Armstrong took legal action against RCA. But Sarnoff had long been prepared for this confrontation, too. An army of RCA's patent lawyers, just like AT&T's lawyers before them, engaged Armstrong and kept dragging the case along in order to consume Armstrong's slim remaining personal resources. With mounting financial frustrations and facing a prolonged legal struggle, Armstrong gradually lost his mental health and patience. Bitter resentment and depression took hold of him, which led to instability in his home. He and his wife began to argue, and on Thanksgiving Day, 1953, after a heated exchange that grew into a physical assault, his wife walked out on him, leading Armstrong's precarious mental state to further destabilize. Two months later, on the evening of January 31st, 1954, Armstrong put on his best formal suit and jumped to his death from the window of his New York apartment. He was sixty-four years old. His widow continued the litigation against RCA and twenty other companies who all used FM technology in their products, at last achieving a bittersweet victory over RCA in 1954, and then went on to win all the other cases, culminated by the final win against Motorola in 1967.

Modulation and Detection of AM and FM Radio Signals

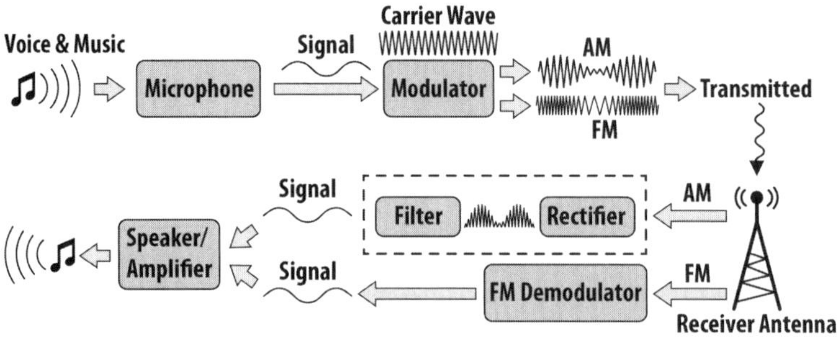

Basic concept of AM and FM radio broadcast. Derek Cheung

The radio signal consists of the high frequency, constant amplitude "carrier wave" and the lower frequency signal which is converted from an acoustic source with a microphone. In AM, the amplitude of the carrier wave is first modulated by the audio signal and then transmitted. The received signal is weaker, but should preserve the shape of the wave. In a radio receiver, the signal is first rectified with a vacuum diode or cat-whisker rectifier, and then the high frequency carrier signal is filtered out using an electrical circuit. The remaining slow varying "envelope" of the amplitude is a replica of the original audio signal, which can be converted back to an audio signal using a speaker. It should be pointed out that if the original radio signal is not rectified, then the positive and negative part of the signal will cancel out, and no signal will be detected.

In FM, the audio signal is encoded as a variation of the carrier frequency using a voltage-to-frequency converter; and the frequency variation is reconstructed in a receiver using the FM decoder. Since frequency variation is less vulnerable to interference from atmospheric conditions than amplitude, the audio quality of FM is superior to AM.

9

TELEVISION

Beginning in the 1920s, radio broadcasting steadily gained in popularity as the public became enamored of listening to music and up-to-the-minute news in the comfort of their homes. At about this same time, Hollywood movies began entering peoples' daily lives, leading people to wonder whether it might be possible to transmit movie-like video images through the air, just like music was transmitted via radio broadcasting. Technologically speaking, this was clearly an extremely difficult challenge, and many considered it a dream that was before its time. However, a few exceptional people took it upon themselves to try to bring those dreams to reality.

TRANSMITTING VIDEO THROUGH THE AIR

Movies and video are not actually continuous moving images. Instead, they are series of slightly different still images, or frames, projected sequentially at a rapid rate. The slow response of eyes to these changes causes people to perceive these images as continuous movement.

On its most basic visual level, each individual frame consists of a two dimensional matrix of spots of light (pixels) of varying levels of intensity and color. Researchers recognized that if every spot of light on each static image could be captured, encoded, transmitted, and then reassembled and displayed in its correct position and sequence, it should be possible to send and receive moving videos. However, for this to happen, several technological obstacles had to be overcome. First, how could a two-dimensional light image be converted into a one-dimensional, serial electronic signal ready for transmission? Second, after receiving that signal, how could a receiver immediately convert and reassemble it back into its original two-dimensional image? Third and finally, how could this process be repeated quickly and continuously, so that all a viewer would see would be smooth, moving video?

The initial hurdle—converting a two-dimensional light signal into a one-dimensional electronic signal—was first successfully addressed by a young German engineer named Paul Gottlieb Nipkow before the concept of television was even conceived. In 1884, Nipkow invented an ingenious optical device consisting of a disk with many round holes arranged in a spiral pattern. In operation, this disk was rapidly rotated to create an optical scan of a two-dimensional space line by line. This process was known as "raster scanning." The detected light from each scanned spot in each image was converted into an electrical signal through a selenium-based, photosensitive detector. Nipkow's device was capable of turning any two-dimensional image into a serial electronic signal, which, in theory, could then be used to reproduce the original still image.

John Baird with his mechanical television setup. Sheila Terry/Science Source

Over the subsequent decades, a number of inventors employed the Nipkow Disk to capture still images. In 1923, a Scotsman named John Logie Baird became the first to successfully integrate and apply Nipkow's electromechanical scanning technology to capture a short video of a moving silhouette. In his early attempts at filming visibly recognizable people, he was forced to use the head of a ventriloquist's dummy as a stand-in for a person—because of the low light sensitivity of the photo-detector, the incandescent lamps required to illuminate the scene were

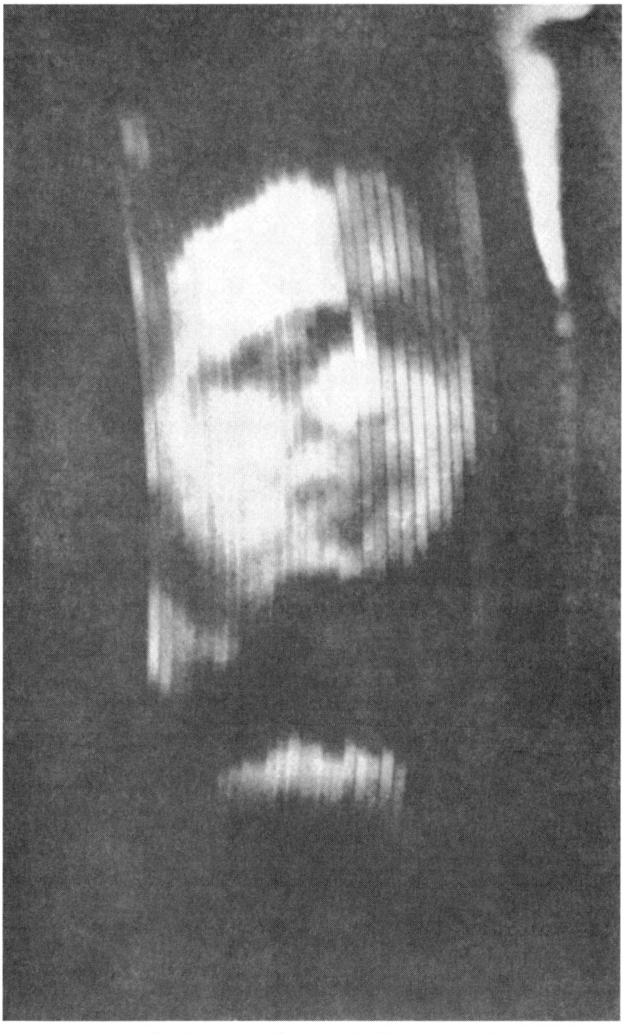

First transmitted television picture. Sheila Terry/Science Source

so bright that the heat they generated was more intense than any living person could stand. But by 1926, after years of work to improve the sensitivity of his photo-detectors (thus allowing him to lower the intensity of his stage lighting), Baird was able to successfully capture the moving image of a nearby office worker. He then transmitted the electrical signal through a wire, converted it back into optical beams, and scanned and projected the original image on a screen—an end-to-end demonstration of the entire television concept using real people as object!

Gleeful, Baird got permission to showcase this device in a London department store. Even though the video displayed was extremely coarse and choppy—only thirty vertically scanned lines, refreshing at the rate of five frames per second—it nevertheless marked the first time in history a real-time moving video of a person was transmitted and displayed. There was no question that Baird had achieved a great success: he had invented the first electro-mechanical television!

After Baird's breakthrough, many major U.S. companies, including GE and AT&T, launched their own in-house television development efforts. Even though the Nipkow Disk played a critical role in bringing the idea of television to reality, its mechanical nature imposed severe limitations on the burgeoning technology, just as Fessenden found himself severely restricted when using a mechanical alternator to generate carrier waves for radio broadcasting. The megalithic corporations did not recognize this truth, however, and they expended prodigious amounts of manpower and material resources trying to improve the Nipkow Disk. Of course, they were traipsing down a path that was essentially a dead end. As with radio, the real key to unlocking the power of television lay in finding a new, electronic way to overcome the mechanical bottleneck. And that key, as it turned out, lay in wait for an open-minded scientist to discover it buried deep in the past.

A FARM BOY FROM UTAH AND A RUSSIAN ÉMIGRÉ

In 1897, Karl Ferdinand Braun invented the cathode ray tube, or CRT. The CRT used an external magnetic field to control the movement of the cathode ray—an electron beam—inside a vacuum enclosure. This magnetic field could both focus and scan the beam, which meant it could be adopted for raster scanning.

As Crookes demonstrated with his Maltese cross experiment, when an electron beam hits the fluorescent screen at the end of a CRT, it gives off light. The intensity of this emitted light is determined by the instantaneous magnitude of the electron beam flux, or in simpler terms, the strength of the signal current. As electrons are unbelievably light, having just 1/1837th the mass of a single atom of hydrogen, or $9.1x10^{-28}$ grams, cathode ray beams can be scanned rapidly with essentially no inertia. Thus, no mechanical scanning system could ever hope to match a CRT for speed and precision in directing a moving beam.

In the Utah countryside in 1921—the same year that Albert Einstein won a Nobel Prize for elucidating the principles behind the photoelectric effect—a fifteen-year-old high school student named Philo Farnsworth became obsessed with the idea of television. He diligently studied all of the information available to him that was related to the technology, which, at the time, was not much. Young Farnsworth did come across an interesting article published in *Nature* magazine in 1908 by a Scotsman named A. A. Campbell Swinton, in which he

Philo Farnsworth. The LIFE Picture Collection via Getty Images

proposed using an electron beam device for scanning and displaying images. Unfortunately, Swinton never did anything to follow up with his idea.

Farnsworth continued his effort to conceptualize his ideas. He claimed that one day, when he was gazing from a small hill at a rectangular wheat field below, he saw a tractor going back and forth as it plowed. This inspired him to think more deeply about using raster scanning to capture and display an image. Thanks to the courses he'd taken at the local college, he knew quite a bit about the raster-scanning properties of Braun's CRT and the photoelectric properties of the cathode first observed by J. J. Thomson, which he correctly recognized to be suitable for converting optical images into electronic signals. Showing tremendous instinct, Farnsworth realized that the then-obscure CRT technology was a natural match to Swinton's idea for making television a reality.

Convinced of his ideas, the fearless Farnsworth, then just eighteen years old, decided to design an all-electronic television system. The system he envisioned would include both a television camera based on the photoelectric effect, which could capture a video image and turn it into a serial electronic signal current, and an image display device with raster scanning, all of which would be constructed with the cathode ray tube technology as its backbone. That he was pursuing a method ignored by industry titans like GE and AT&T did not dissuade him in the least.

With the help of his friends, Farnsworth raised $6,000 to launch his entrepreneurial effort. At first, he took his CRT tube design to a well-known glassblowing shop in Salt Lake City to have it built, but he was not satisfied with the quality of the work. Full of the exuberance of youth, Farnsworth decided to learn glassblowing techniques so that he could construct his vision himself. After achieving some success, his reputation grew, and Farnsworth was able to raise another $25,000 from a wealthy San Francisco banker. At that stage, Farnsworth realized it was time to leave the farm and move to the city. He set up a laboratory near downtown San Francisco and continued to develop and refine his invention.

In 1927, when he was only twenty-one, Farnsworth successfully completed the first prototype of his all-electronic television camera and video display device, making him one of the youngest major inventors in all of human history. Along the way, he applied for and obtained a number of important patents. In 1928, Farnsworth officially unveiled his refined television system in public. The very first image on his CRT screen was a "$" sign—a subtle dig at his investors' near-endless complaints about the outlays required to finance his research and a thumb in the eye of GE and AT&T, who had spent far more money developing their far less technologically advanced systems. Although Farnsworth's television technology was still in its infancy, this one man working nearly alone was able to produce a device whose functionality was already superior to those being

Vladimir Zworykin. Courtesy of Hagley Museum and Library

produced by teams of experts at several of the largest companies in the world. His exhibition proved that CRT technology was the right choice for television—this country boy from Utah was on the right track!

As it happened, Farnsworth was not alone in attempting to develop television technology based on the CRT. A Westinghouse employee named Vladimir Zworykin was also deeply intrigued by the prospects. Zworykin, a young researcher who fled to the United States in 1922 to escape from the Russian Civil War, was hired by Westinghouse to design new electrical power products. However, he had been studying the properties of the CRT for almost a decade before his exodus from his homeland, and his mind was always drawn back to the idea of using the CRT technology for television. At every opportunity, he lobbied Westinghouse management to allow him to start a research project into television technology.

In early 1923, Zworykin's persistence paid off. Westinghouse management tacitly consented to his doing some small-scale experiments in addition to performing his existing duties. After some work, Zworykin was able to apply for a patent based on his early thoughts about a CRT-based television. Two years later, however, management demanded a review of his project, and even though Zworykin had made some technical progress, many of the original concepts contained in his patent proved flawed, needing major improvements in order to have any hope of success. Seeing this television development effort as a diversion from their core business, Westinghouse management decided not to continue funding the project any longer, instead transferring Zworykin to a new company in which they owned a 20 percent stake: RCA.

As it happened, the move was a great fit. RCA's radio and broadcasting business was growing rapidly at the time, and Sarnoff was himself scouring the worlds of electronics and broadcasting for the next growth opportunity beyond radio. Upgrading from a voice and music media to a lifelike video experience was a natural extension. Sarnoff, a fellow Russian émigré, liked what he heard from Zworykin, and he agreed to support Zworykin's efforts to develop a CRT-based television. Sarnoff did not even blink when Zworykin requested $100,000 in funding to complete his television project, not knowing that this sum was four times the amount that Farnsworth had needed to build his own CRT system. Sarnoff's backing of Zworykin's wager would mark an unprecedented level of investment in a new technology by a private sector company—Sarnoff certainly had no idea at this juncture that, by the time RCA officially made its first television broadcast a full decade later, its total investment would come to exceed $50 million, or 500 times the figure to which he'd originally acceded!

In 1929, more than one year after Farnsworth's initial success, Zworykin and his team at RCA were finally able to demonstrate a complete, working television system. The RCA and Farnsworth systems shared many similar features, but RCA's television camera was more sensitive and offered better picture quality. This was not surprising, since Farnsworth was essentially competing single-handedly against a team of technical experts fully backed by the resources of a large corporation. It also didn't hurt that Zworykin and his collaborators visited Farnsworth's laboratory while he was away on a trip to glean working knowledge of his products a full year before demonstrating their own prototype. Did they copy any of Farnsworth's designs? Zworykin did not say.

At Sarnoff's behest, Zworykin also integrated FM technology into the RCA television to serve as its sound transmission and reception system. This, as it happens, was Sarnoff's plan all along, and it partly explains why he was reluctant to engage Edwin Armstrong in serious discussion about purchasing his patents for the FM radio. FM broadcasting technology provided a solution for audio transmission for television, and it also had the added bonus of creating a rationale for the FCC to reallocate the FM frequency band, crushing the ambitions and vision of Sarnoff's principal competitor.

THE INTELLECTUAL PROPERTY BATTLE

Philo Farnsworth was the first to demonstrate the revolutionary CRT-based television system, an impressive achievement made doubly so by his improbable background. His indomitable spirit truly was remarkable. However, huge amounts of resources were required to fully develop this revolutionary technology and push it all the way to market creation and mass production—far more than what Farnsworth and his investors could afford.

Sarnoff was determined that RCA dominate the television market just as it had with radio. However, he knew well that numerous basic patents held by Farnsworth posed a significant threat to his plan. His strategy was to throw around the weight of RCA to paint Farnsworth into a financial and legal corner, just as he'd done with Armstrong. In preparation for this showdown, RCA quietly acquired a significant portfolio of television-related patents from other inventors, mostly from Europe. With a full team of legal advisors, Sarnoff prepared to squeeze Farnsworth out of the television picture and crush the young inventor in court.

In 1933, the legal battle began. Farnsworth's technological achievements and precedent-setting made the foundation for his case almost unshakable. Even so, the RCA lawyers tried every legal maneuver at their disposal to discredit, pressure, and drain the solitary inventor. Farnsworth, just as Armstrong before him, simply did not have the resilience and strength to confront RCA's overwhelming advantages in resources and influence. At the end, Farnsworth succumbed, selling all thirty-six of his television patents to RCA for a mere one million dollars. RCA even pressed Farnsworth into agreeing to receive payments in ten equal installments for ten years without interest. It was not an insignificant sum by any stretch, and surely, Farnsworth emerged from his battle with RCA far better than Armstrong had, but the final price he received from Sarnoff was still shockingly meager given the scope and value of what he was forced to concede.

Like Armstrong before him, Farnsworth left his legal battle disheartened and at the brink of depression. Over the following years, he squandered his money in pursuit of a questionable approach for energy generation based on nuclear fusion. Like Tesla, he became yet another talented pioneer caught chasing an unobtainable dream after his initial success had peaked and ebbed.

With Farnsworth out of the picture, the RCA public relations machine trumpeted the corporation's contributions to television technology as loudly as possible. Based on the patent he filed in 1923 while he was still a Westinghouse employee, RCA falsely claimed that Zworykin was the first and sole creator of television, sidestepping entirely the fact that the ideas contained in Zworykin's original patent never actually worked! Farnsworth remained a largely forgotten figure—as is often the case, the victors have the advantage of writing history.

By 1939, RCA was ready to introduce the world to the television era. Sarnoff picked the opening day of World's Fair, held in New York, for the historic unveiling. As President Franklin D. Roosevelt delivered his opening speech, real-time black and white video images and sound were captured and broadcast to over twenty television receiving sets distributed throughout the city. All across New York, awed citizens far from the fairgrounds witnessed their President ushering in a new age in mass communications.

Though television was officially launched at the World's Fair in 1939, its widespread rollout was delayed by the Second World War. Despite this postponement, television rapidly became a great commercial success. At the end of 1946, the first year that television was commercially available, RCA sold 170,000 sets priced at $375 each—the equivalent of $3,340 in inflation-

History's first live television broadcast in 1939. Courtesy of Hagley Museum and Library

adjusted 2010 dollars. However, by 1948, the total number of television sets sold had increased six-fold, to over one million. The live broadcast of that year's World Series between the Cleveland Indians and Boston Braves, a television first, assuredly helped raise national interest. By 1950, the number of television sets sold had grown multiple times over again, reaching seven million. RCA controlled a 50 percent share of the television market; in addition, other manufacturers were obliged to pay significant royalties to RCA when selling their own television products. RCA's initial investment of $50 million was recouped within a few short years, and CRT technology dominated the television market until the early years of the twenty-first century.

As television gained in popularity, the center of gravity of broadcasting media gradually shifted from radio to television. Watching television became an indispensable activity in people's daily lives—and, as the dominant player in the industry, RCA enjoyed tremendous business success throughout the 1950s and 1960s. In 1970, Sarnoff retired and passed the reins of the company to his son. Without David Sarnoff, however, RCA lost its business vision and direction. It was mismanaged, squandering valuable resources in pursuing other, unrelated businesses, such as car rentals and frozen TV dinners, and it failed in its attempt to compete in the emerging computer industry. In 1984, the failing RCA was acquired by GE, ironically, the very company that first started RCA in 1919 under the encouragement of the U.S. government. By the time of its buyout, the main value of RCA was its NBC network.

Even though commercial television broadcasting made its debut in 1939, World War II forced a seven-year lag until its widespread rollout in 1946. During the war period, the RCA team continued to make improvements to CRT technology, of course—just not for use in television. Instead, the CRT technology found another application that was far more pivotal to the battle for the American way of life: radar.

10

RADAR

In the late 1930s, the United States basked in the glow of peace and prosperity. The fog of the Depression had lifted, and new forms of entertainment, such as radio broadcasting and Hollywood movies, were sweeping the nation. Television was also on the cusp of being launched. But in Europe, storm clouds were gathering as the Allied and Axis countries prepared for another global battle.

During the lead-up to war, an official in Great Britain's Ministry of Defense read about Tesla's "Wireless Death Ray" idea and found it intriguing. He also received intelligence that the Germans might be developing such a secret weapon. Concerned, the official wrote a letter to Britain's authority on wireless technology: Robert Watson-Watt, a descendent of James Watt, inventor of the steam engine. Watson-Watt wrote back a candid reply based on detailed scientific analysis, pointing out flaws in the assumptions Tesla used in the death ray concept. He concluded that a beam weapon based on electromagnetic waves was not physically realistic. However, he noted that electromagnetic waves could be used to detect distant targets such as airplanes and warships. In addition, they should be capable of deducing any object's distance by analyzing the time intervals between wave emission and the reception of the signals reflected back from the distant object.

Watson-Watt included a basic sketch for such a system. The concept he elucidated would later be called Radio Detection And Ranging, but it was better known by its acronym: RADAR.

Robert Watson-Watt. © Hulton-Deutsch Collection/ CORBIS

CLAIRVOYANCE

Though Watson-Watt's letter was influential, radar was not in fact a new concept. Back in the 1880s, Heinrich Hertz proved in his experiments that radio waves could be reflected by a sheet of metal, just as light is reflected off of a mirror. Using this principle, a German engineer in 1904 first demonstrated a rudimentary system that could detect a boat in heavy fog, even from as far as five miles away. By the early 1930s, many countries had begun to fund small-scale radar development efforts. Still, it was not until the British military moved forward with Watson-Watt's proposal that anyone even imagined the full breadth of the power of radar technology—or its full importance.

In 1935, the British government approved a proposal based on Watson-Watt's concept and began to design and construct a top-secret radar system, code named Chain Home. It consisted of multiple clusters of radar sites situated along the southeastern coast of England. The intent of the system was to allow for early detection of airborne incursions by German war planes. Though Europe was not yet at full-scale war, the German Luftwaffe was gaining in strength and size, and Hitler's increasingly aggressive intent was obvious.

The technical challenges of building a working radar system were formidable. For instance, in order to detect a remote target of unknown position, a broad area would need to be blanketed by electromagnetic waves, only a very small portion of which would bounce back from the intended target. Those signals would therefore have to travel round-trip between the transmission station and the object. Since signal power decreases rapidly with distance, a high-performing radar system would require an enormous amount of transmitting power, as well as extremely high-sensitivity receivers. To increase spatial resolution and diminish background clutter (that is, unwanted signals bounced back from other sources that degraded the intended target signal), it would also be desirable to construct a method by which to focus and scan the beam of the electromagnetic waves transmitted, just like an optical searchlight.

Watson-Watt explained that the ability to focus the electromagnetic beam would be determined by the wavelength (or frequency) of the radar wave and the size of the transmission antenna. For small- to modestly-sized antennae, the wavelength of the radar radiation would have to be one meter or less, corresponding to a frequency range at least ten times higher than those of radio waves, in a spectrum known as microwaves. Unfortunately, there existed at the time no technology that could successfully generate these high-power microwaves.

Watson-Watt understood the science behind creating a radar system. But charged now with developing the system not in theory but in practice, he found himself at an impasse: there was no way to generate the microwaves that a reasonably sized antenna would require. What's more, war seemed more and more of a certainty, leaving no time to wait for the development of a microwave power source for Chain Home. The only choice was for Watson-Watt to create a workaround. His solution was to use high power radio waves with wavelengths in the 12-meter range (20–50 MHz), which could readily be created by specially engineered triode oscillators. The long wavelengths of these radio waves limited their spatial resolution, so to compensate, Watson-Watt scaled up the height of his transmission antennae to an incredible 360 feet.

Thanks to Watson-Watt's leadership, the Chain Home system was completed on time. The radar beams were not scanned, but instead operated in a crude "flood illumination" mode, aimed at an altitude over 500 feet to reduce ground-based interference. Even though the system was designed and built using an inelegant brute force approach, it worked. In early test runs, it demonstrated the ability to detect German aircraft grouping over France as far as 120 miles away from the English coast. And even though the resolution of the radar used in the Chain Home system was limited, unable to determine the exact number of German aircraft or the precise direction of their approach, it still provided ample early warning of invasive forces and their approximate arrival time and location. This provided enormous advantage to the British Royal Air Force, which could wait and ambush incoming Luftwaffe rather than be caught by surprise. This clairvoyance in coastal defense was critical, particularly during the 1940 Battle of Britain. In that skirmish, the Luftwaffe suffered heavy losses

Chain Home radar transmitting tower. © Bettmann/CORBIS

and remained unable to establish air superiority over the English Channel, quashing Hitler's plan to invade England altogether. The power of radar was pivotally important to the British RAF's victory.

HUNTING THE SUBMARINE

By late summer, 1940, radar seemed to have given the British some hope of saving themselves from the Nazi war machine. Nonetheless, the attitude in England remained tense, particularly because of the German U-boats. These wolf packs of submersibles frequently savaged the Allies' North Atlantic convoys, causing severe losses of troops and supplies and creating enormous difficulties in maintaining a supply line to the European front. Unfortunately, British sea power was not primed for underwater battle, which led British military planners to wonder: if it worked so well in turning away aerial attacks, wouldn't radar be useful in detecting the presence of underwater German craft as well? If that technology could be harnessed for naval purposes, even submarines camped in the vast reaches of the ocean would be impossible to hide.

Unfortunately, radar doesn't work underwater. But the British became convinced that one effective way to thwart the growing scourge of U-boat attacks would be to use high-sensitivity radar to detect German U-boats when they surfaced. If radar systems could be mounted on a large fleet of aircraft patrolling the skies, that would give the British full view of the seas. As soon as a submarine surfaced, even if with only its periscope, it could be discovered and dealt with. However, any antenna system built onboard an aircraft would have to be compact and light; war planes could not carry 360-foot tall antenna structures. No easy workaround was possible here. There was no alternative. The Allies would have to develop a high-powered, compact, microwave power source if they were to have any chance of retaking the seas in their bid to win the war.

With the fate of many nations in the balance, the British defense establishment mobilized, turning to the country's science and engineering community for a solution. Responding to the challenge, John Randall, a professor at Birmingham University, and his graduate student, Henry Boot, carefully examined

John Randall (left) and Harry Boot with the first cavity-magnetron (taken in 1975). Photo Duffy © Duffy Archive

every point of data they could find regarding microwave energy sources. Combing the whole known world of microwave technology, they found that two existing vacuum tube devices, the *magnetron* and the *synchrotron*, could produce microwaves of the required frequency range. These devices both operated via the interaction of electron beams with magnetic fields in a vacuum. However, these devices were also both woefully underpowered, capable of producing only five to six watts of power at maximum. In spite of these limitations, Randall and Boot believed that with further improvement, these technologies might offer the solution to the RAF's needs.

By analyzing the existing devices and gleaning insight from the theoretical works of Maxwell and Hertz, Randall and Boot hit upon the concept of equipping the magnetron with a resonant cavity to increase its output power. A resonant cavity is a space in which microwaves can regeneratively amplify themselves, analogous to Armstrong's feedback loop. Randall and Boot didn't have much time to perform detailed mathematical analysis of their concept or to compare and contrast potential design tradeoffs. Every day and night, Allied transport ships were sunk in the depths of the North Atlantic, sending countless young sailors to their watery graves. They had to act with great urgency. Working as swiftly as they could, the two quickly sketched out their new design, augmenting a nearly finished ordinary magnetron rather than constructing a new device from scratch. Within a few weeks, the modified magnetron was finished, with the resonator cavity elegantly integrated into the device's structure. Holding their breath, they tested it—and something amazing occurred.

Previous magnetrons had been capable of producing microwave radiation of the frequency needed—about 3,000 MHz, or 3 GHz—but only at the relatively low power of five to six watts. Randall and Boot's first resonator-equipped magnetron produced electromagnetic waves at *600* watts—a level of power one hundred times greater than the highest limit ever previously reached! Such a remarkable level of improvement is quite rare in the history of technological advancement, but it demonstrates a simple truth: armed with a solid understanding of fundamental science and innovative engineering design, one may be able to realize the full potential of a new technology.

After a few more weeks of fine tuning, Randall and Boot were proud to unveil a modified magnetron that could be held in one hand but was capable of producing 1,000 watts (1 kW) of continuous power, with a peak output of up to ten kilowatts when operated in pulsed mode. At this power level, an airborne radar system could theoretically detect the periscope of a U-boat up to seven miles away. Best of all, thanks to the high operating frequency of these devices, their radar antennae were small enough to be easily fitted onto aircraft.

At first, Randall and Boot were ecstatic—but then, they ran into another formidable technical problem which was related to the fundamental principle of magnetrons. The frequency of the microwaves generated by the magnetron was unstable; it kept drifting during the course of operation. This would be analogous to a radio station broadcasting at a random, continuously changing spot on the dial, forcing listeners to constantly re-tune quickly throughout the course of a song. This was a major show-stopper, but Randall and Boot already had a solution in mind. They were able to create a clever feedback sensing circuit that would dynamically and automatically adjust the radar receiver's "listening" frequency in lockstep with the variations in the frequency of the transmitted wave. Their workaround was successful. During real-world trials several weeks later, Randall and Boot's system was able to detect bicycle riders on a country road ten miles away from the test setup!

The success enjoyed by Randall and Boot in creating the powerful magnetron was nothing short of remarkable. Still more remarkable, though, is that this Herculean effort took only nine months from start to finish, from conceiving of the resonant cavity to successfully demonstrating the microwave radar in operation. It was precisely the kind of extraordinary event that could only happen under the extreme urgency of war. Speaking of the war, the Germans were never able to match the range performance of their airborne radar with the Allies, so they were at a disadvantage throughout the war. It turned out that the German engineers decided early on to pursue the klystron technology, rather than the magnetron technology; and they had not come up with an equivalent resonant cavity concept to boost the klystron power. The brilliance of technical minds did make a huge difference on the outcome on the battlefield.

THE MOST VALUABLE LUGGAGE

Although the technology critical to producing airborne radar was now in hand, developing, manufacturing, and deploying a full range of radar systems on aircraft remained a formidable challenge. Great Britain found itself completely consumed by the war effort; the country was already expending its full manufacturing capacity to produce planes, guns, bullets, shoes, food, and other necessities. Developing and pushing a family of new radar systems into mass production was, at the time, simply unfeasible.

Facing a shortfall in human resources and production capability, Prime Minister Churchill resolved to seek outside help. In September 1940, a secret British delegation led by senior scientist/administrator Henry Tizard arrived in the United States. With him, Tizard carried a special piece of luggage in a wooden box: a short, copper cylinder, drilled through with a complex set of holes. To the untrained eye, it had little to no value whatsoever. But it was this metal object that changed the balance of the war. This little box may have been history's most valuable luggage.

In return for shared ownership equity in the new magnetron technology, the United States agreed to assist the British in jointly developing and rapidly mass producing an airborne radar system for deployment in the European theater. The War Department set up a dedicated technical organization, appropriately named Radiation Laboratory, or Rad-Lab, on the campus of MIT in Cambridge. Rad-Lab was staffed with the most outstanding technical talent from across the entire United States, including electrical engineers, physicists, and mathematicians.

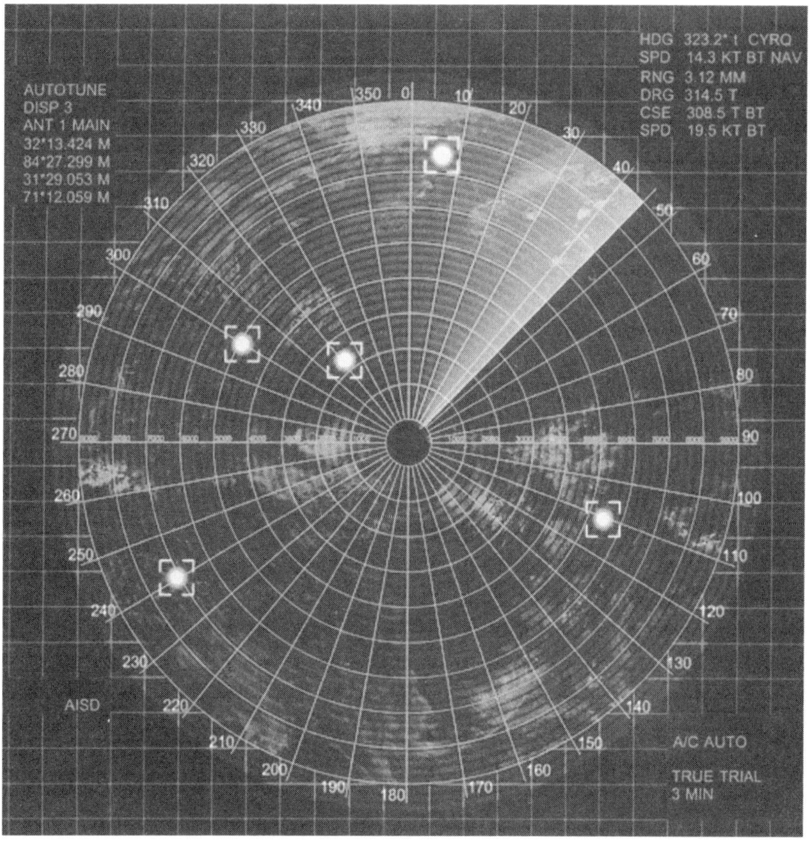

CRT screen used for radar signal monitoring. Andrey Prokhorov/ Getty Images

Some came from America's most prestigious academic institutions, while others had been in the research departments of top high-tech private industries. All were in awe of the output power of Britain's resonant-cavity magnetron, and all were determined to leverage this breakthrough to help the Allies win the war.

It was a vast assemblage of incredible talent, and each member worked together under a sense of extreme urgency. Though the United States had not yet officially entered the war, the writing was on the wall, and everyone was fully dedicated to quickly developing a wide range of radar systems that could be used in battle. Work was organized into teams: some focused on perfecting the magnetron for mass production, others designed new antennae and supersensitive microwave receivers, and still others collaborated on developing highly efficient algorithms for processing radar signals. One unique challenge was determining the proper way to display rapidly changing target information within the overall threat environment. The solution was to use a green-hued CRT to show and constantly update the locations of detected targets. Thus was CRT technology refined during the war, even though commercial application of the television had not yet commenced. Within a year—a year at the close of which the Japanese bombed Pearl Harbor, pulling the United States definitively, unquestionably, and openly into the war on the side of the Allied Forces—several families of high-performance radar systems had been developed for various battlefield uses, including airborne submarine surveillance.

The War Department issued production orders to companies across the country, and within months, radar systems were installed on U.S. and UK aircraft patrolling the North Atlantic. Sure enough, airborne radar, coupled with the development of depth charges with sonar tracking sensors and the successful breaking of German codes by British intelligence, led to large swaths of German U-boats becoming the casualties of Allied attacks. By the spring of 1942, very few German U-boats were still prowling the North Atlantic. Convoys of transport ships were able to sail safely from North America to England, bringing a steady supply of goods and material to Great Britain that would lead to the massing of forces for the Normandy invasion. Through the end of the war, the German navy could not fully understand how or why the tide had turned so abruptly against its submarines.

Airborne radar had multiple applications. In addition to submarine and surface ship detection, it was used to aid in high-precision bombing. In aerial combat, too, its ability to spot the first signs of the enemy aircraft, whether in night or day, made for better tactical coordination in achieving air superiority. But as important as radar was for the air force, it proved even more essential to the navy.

For thousands of years, sailors had to stand on the masts of their ships and use their naked eyes, or later, telescopes, to scan for land or the sight of enemy vessels. However, thanks to radar, those days were gone forever. On the vast oceans, no matter whether in dense fog or the pitch-black of night, shipboard radar systems could clearly detect shorelines, as well as every object in nearby waters. As the range and precision of coverage grew, onboard radar systems became indispensable to all ships. This gave the Allied naval forces a crucial advantage. After the war, the technology broke past the borders of military use, and nearly all of today's oceangoing vessels—even civilian leisure boats—have several different sets of radar on board, each playing its own crucial role.

RADIO NAVIGATION

Another related technology that benefited from the rapid wartime development of radar was radio navigation. Within Watson-Watt's radar development team worked a legendary wireless expert named Robert Dippy. Based on the well-known triangulation principle, Dippy proposed the concept of using multiple stations to transmit precisely time-synchronized radio signals. These signals would necessarily arrive at any given target location at slightly different times, due to the differences in distance between the signals' origin points and their common destination. If the location of every transmitting station was known precisely in advance, the exact position of the distant object could be derived based on the difference in signal arrival times.

Robert Dippy. Courtesy of Dr. Phil Judkins, Purbeck Radar Museum, and Mr. Bob Fisher, radararchive.com

Even though the principle of triangulation was simple, the challenge to build a practical radio navigation system with the desired accuracy was extremely difficult. Dippy cleverly borrowed and modified many cutting-edge radar and radio communication technologies to the navigation project, and in 1938, his team demonstrated the feasibility of a practical system. Beginning in 1939, Dippy and his team began to build the first operational radio navigation system, code named "GEE." At the end of 1942, the British air force successfully used GEE for navigation in its night bombing mission of the tactically valuable German industrial city of Essen. Even though the accuracy over long distance was limited to about one kilometer, it was adequate to successfully guide the planes to the vicinity of the target area, and then navigate them safety back to the home field. The GEE system was further refined and played a pivotal role during the Normandy invasion in 1944, when the physical position of tens of thousands of moving platforms of the Allied forces had to be closely coordinated across air, land, and sea.

During his mission to the United States in 1940, Tizard offered to share Britain's knowledge of radio navigation technology along with the magnetron in exchange for U.S. assistance. So concurrent with the radar development effort, the Rad-Lab established a parallel effort to investigate Robert Dippy's radio navigation technology. This team worked equally brilliantly as the radar team, and they eventually succeeded in developing a modified navigation system known as LORAN, for LOng-range RAdio Navigation, which offered an extended operating range of up to 1,500 miles.

During the U.S. Navy's struggle to push back the Japanese, they often launched air combat missions from aircraft carriers, as they had few forward land bases from which to incite counterattacks. However, planes taking off from aircraft carriers typically had small fuel tanks in order to ensure they could achieve liftoff from a short runway. Carrier-based aircraft also had to keep fuel weight low in order to increase their maximum payload, or the amount of ordnance they had on board. But the Pacific Ocean was vast, with extensive cloud coverage, and even in clear weather conditions, there remained very few visible landmarks for navigation. Even slight navigation errors might lead pilots to be in danger of not having enough fuel to return to their carrier ships. Thus, under such circumstances, radio navigation proved all the more indispensable. LORAN was a critical system capability that allowed U.S. naval pilots to successfully fight in the Pacific.

Years after the war, the LORAN system was upgraded to LORAN-C, which, like radar in ships, became widely used in all civilian aircraft. In the 1980s, radio navigation took a major step forward when radio transmission stations were installed on a constellation of orbiting satellites that covered the entire globe. Each satellite was equipped with ultra-high-precision atomic clocks to synchronize navigation signals, and their positions were precisely tracked by ground-based radar. That newest system, which came to be adopted in everything from ships to cars to cell phones, became known as the Global Positioning System, or as we refer to it today, GPS. This man-made system provides global location service with accuracies and capabilities way beyond compasses based on detecting earth's natural magnetic field, which was the first application by human beings of the electromagnetic phenomenon.

THE MICROWAVE WORLD

During the war, the total amount of funding devoted to the development of radar was second only to that allocated to the Manhattan Project, which created the atomic bomb. Such a massive scale of investment dwarfed anything ever before seen in the history of electronic technology, including the $50 million invested by RCA in television. More importantly, the technical teams assembled for radar development provided a boundless source of well-trained and talented scientists and engineers who were capable of solving increasingly complex and multidisciplinary technical problems. Such an outlay of money and expertise led to the discovery of a vast array of new applications that continued to play important roles well after the end of the war.

One of the elements key to the success of radar was the dramatic improvement in high-power microwave power generation. Thanks to Randall and Boot, a small, compact magnetron could generate over 1,000 watts of microwave power. In the device's early days of development, technicians in the laboratory noticed that they could easily light the tip of a cigarette by placing it near the output horn of the magnetron. They also found that they could dry wet clothes quickly by putting them in front of the magnetron, and that chocolate bars in their pockets melted when they walked into the lab. The property of microwave energy to turn into heat when absorbed was eventually leveraged to develop a practical application: the microwave oven.

As the demand for radar products grew in the decades that followed the war, the production cost of magnetrons commensurately dropped, and the microwave oven became an affordable and then an indispensable kitchen appliance—an application probably never expected by Randall and Boot. Magnetrons convert almost half of their electric power input (that is, from a plug-in wall socket) into microwave energy, and in a microwave oven, almost all of the microwave energy created is absorbed by polarized molecules, such as water molecules in food and converted into heat through a process known as dielectric heating. Thanks to this process, microwave ovens are in fact the single most energy-efficient cooking appliance.

During the war, massive research efforts were devoted to shortening the wavelengths of microwave emissions so that radar-employing craft could use ever smaller antennae while achieving higher spatial resolution of targets. However, engineers noticed that at some specific wavelengths, the propagation range for these radar waves dropped precipitously, a phenomenon no one could explain. During the war, there was no time to investigate such intellectual questions. But after the war ended, many professors who were working on radar development returned to academia, and there they found rich opportunity for further research. What they eventually discovered was that at very specific frequencies, microwaves become absorbed by molecules of gases in the atmosphere, including oxygen, nitrogen, water vapor, carbon dioxide, ozone, and many others. These gases selectively absorb only specific wavelengths of electromagnetic radiation based on the resonance energy of the molecular structure of each unique gas. In other words, every gas has its unique spectral "signature" of microwaves that it exclusively absorbs. This finding gradually evolved into the science of microwave spectroscopy, which has become an important tool in modern chemistry and physics as well as atmospheric science. Using microwave spectroscopy, scientists can measure various phenomena in the atmosphere, including the levels and types of air pollution and distribution of carbon emissions.

Radar found still other uses in the postwar era. It was discovered that when microwaves bounce off a moving target, the frequency of the reflected electromagnetic wave changes slightly, a phenomenon known as Doppler shift. The speed and direction of the target's movement can be measured from this small shift. Highway patrol officers use this effect to determine the speed of moving vehicles and to give out dreaded speeding tickets. Mounted in automobiles and coupled with cruise control, this same technology has been employed in assisted-steering and braking devices to help avoid collisions, and it may even lead to a potential path toward creating self-driven smart cars. Still another important application lies in weather monitoring. Weather radar utilizes the Doppler Effect to detect the spatial distribution and movement of raindrops of various sizes and densities in air. Images generated by weather radar can show the amount of rainfall at different places, which is indispensable in predicting and reporting weather such as is shown on the television news. Furthermore, the nose cones of many commercial airplanes are fitted with compact weather radar to detect storms and other air turbulence several miles in front of the aircraft so pilots can change flight courses to minimize danger or discomfort to passengers.

After the end of the Second World War, the United States and the Soviet Union confronted each other in a frigid, global standoff that lasted more than forty years: the Cold War. During this period, radar development continued at as breakneck a pace as it had during World War II. The main objective for both sides was to develop a nationwide early warning radar system that could continuously monitor the entire air space for all flying objects, including both bombers and missiles. The project was code-named SAGE, which created many breakthrough advances in radar, computer, and system technology. Though the nuclear holocaust feared by many thankfully never took place, civil aviators later adopted a derivative of this global radar technology for use in air traffic control. Nowadays, the position, speed, and direction of almost every aircraft in the sky are continuously monitored by an immense, integrated radar system. With the aid of powerful computers, GPS, and communication links between pilots and ground control, air travel today is safer and more efficient than ever.

In the postwar period, many physicists and university professors received surplus radar equipment from the U.S. Defense Department. Some used this equipment to study not points on earth, but points in space, beginning with the moon. By increasing microwave power levels and enhancing receiver sensitivity and signal processing capabilities, researchers were able to use radar to directly measure the exact distance between the moon and the surface of the earth. During this process and in other observations, they were surprised to discover numerous weak microwave signals originating from somewhere deep in space. Faraway celestial bodies, it seemed, emitted these signals. Combined with spectroscopy, research in this area led to the establishment of still another new realm of space study known as radio astronomy.

Maxwell was the first to point out that microwaves, radio, and visible light all belong to the same family of electromagnetic radiation; they differ from one another only in wavelength. Max Planck's blackbody theory illustrated that any object above minus 273 degrees Celsius (also known as absolute zero, when all kinetic motion is "frozen") emits some amount of electromagnetic radiation in all wavelengths. The extremely low temperature objects of deep space (and some gases) emit most of their radiation in the microwave frequency range. Radio astronomy made observing planets, stars, and interstellar dust in the microwave spectrum just like looking at them in a color scheme that our eyes can see, and thanks to its development, mankind was able to greatly expand upon its ability to observe and understand the wonders of the universe.

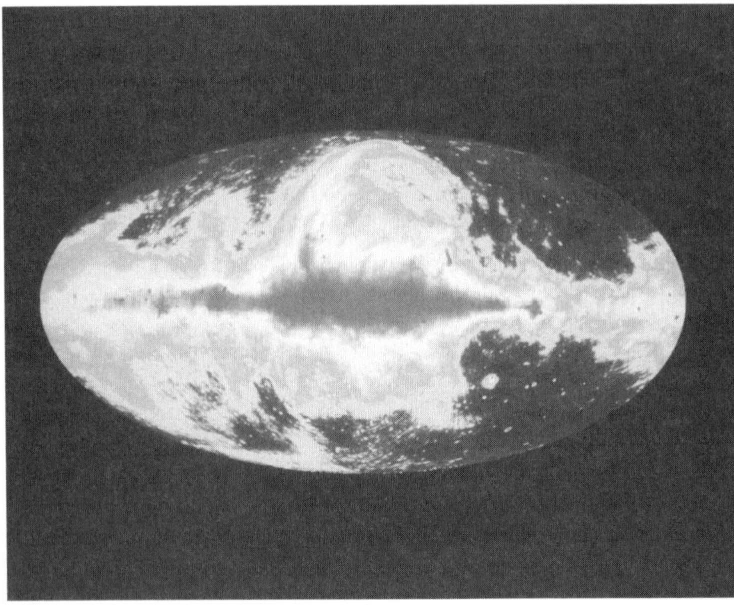

Milky way viewed in radio wave frequency. Max-Planck-Institut für Radioastronomie/Science Source

Through microwave research directed toward outer space, man was able to shed light on many fascinating topics, including the Big Bang Theory, postulated as the origin of our universe. This drove radio astronomers to demand ever-higher levels of receiver sensitivity that would allow humans to detect even the faintest signals from heavenly bodies at the edge of the universe. Engineers and scientists responded by pushing the sensitivity of microwave sensing devices to unprecedented levels, which even planted the seed for the eventual invention of the laser. It is amazing to realize that a large body of our knowledge about the universe can trace its roots to the development of wartime radar.

11

COMPUTER

The ability of the vacuum triode to amplify weak analog signals was quickly leveraged to effectively tackle the long distance telephony challenge. Next came the media revolution that led to the birth of radio, wide-scale broadcasting, and television. With the outbreak of the Second World War, technologies based on electromagnetic waves extended into the fields of radar and radio navigation. And then, in the 1940s, vacuum electronics facilitated yet another major new breakthrough of inestimable importance: the computer.

THE CALCULATING MACHINE

Strictly speaking, a computer is not necessarily an electronic product, but rather a tool to carry out computing and logic operations based on a set of instructions. Over one thousand years ago, the Chinese were already using a beaded device called the abacus to quickly perform arithmetic calculations. At the beginning of the nineteenth century, the English mathematician Charles Babbage designed an ingenious mechanical computing machine called the "Difference Engine," which could rapidly perform a large volume of complex repetitive calculations. It could, for example, generate function tables or analyze census data in a variety of ways. By the late 1930s, mechanical computers of various designs were in wide use in Great Britain, Germany, and the United States.

These first, powerful computing machines were all based on clever mechanical mechanisms. However, with advances in electrical technology, precision electrical motors began to be introduced into a new generation of faster, electromechanical computing machines. By the early years of the twentieth century, many logic and computing functions were rapidly and precisely executed by these electrically assisted mechanical mechanisms. The Mark I computer built at Harvard was the most advanced of its day; it set a record by being able to execute up to three basic addition or subtraction operations per second.

The invention of the vacuum triode opened the door for a new class of computing machines. As elucidated earlier, the vacuum triode offered two important functions: signal amplification and switching. The electronic switch feature was extremely fast and energy efficient, and it was well suited to performing logic and computing operations. In 1919, a British engineer named W. H. Eccles used a vacuum triode and a capacitor to demonstrate a simple "Flip-Flop" circuit, a digital 0-or-1 determinant that became the most basic building block for the operation of binary logic functions and for digital data storage. As with oscillator circuits for radio and CRTs for television, it became obvious that using electronic circuits such as the triode-based flip-flop would allow for the design of a much faster, versatile, and powerful computing machine than any mechanical or electromechanical device might hope to be.

At the outset of the computing age, vacuum triodes were still far too expensive for large-scale use. A computing machine might require from several hundred to a few thousand triode circuits. Given the expense inherent in such a task, building such a vacuum tube–based computing machine appeared to be a fool's errand, not only in terms of cost but also in terms of the technical challenge it presented. However, with the rapid market growth of radio, the price of vacuum tubes began to fall, and at the same time, the tubes' performance and reliability improved. Beginning in the 1930s, several pioneers, notably a German named Konrad Zuse and a Bulgarian-American named John Atanasoff, began small-scale experimentation using vacuum tube circuits in their computing machines.

The prospect of fast and powerful computing machines presented a tantalizing picture to the military. Coupled with special mathematical and logic instructions, electronic computing machines could be used to encrypt messages or break codes, perform high-speed calculations related to the expected trajectory of projectiles, or solve real-time optimization problems regarding complex logistics operations. The technical challenges facing these computing machines were large, always related to how to perform each calculation faster, with more accuracy, and with more flexibility to tackle different types of problems.

During the buildup to war, the British pursued substantial advances in electronic computing technology, not only in hardware design and construction, but also in developing the "intelligence" of such machines, which was pioneered by mathematician Alan Turing. In 1941, unknown to the outside world, and under strict military secrecy, British engineer Tommy Flowers and his team developed the world's first programmable, high-performance electronic computing machine using over 1,500 vacuum triodes. The machine was code-named Colossus, and it played a major role for British intelligence in deciphering and analyzing many encrypted messages used by the German forces. After the war ended, Churchill himself ordered the machine to be cut into pieces so that spies could never leak the top-secret technology to the Soviets. The blueprint for the design was finally released in the 1970s and the machine was re-created by a group of computer history hobbyists in 2007. However, because of the top-secret nature of the project, the ground-breaking technology embodied in Colossus had little direct influence on the subsequent development of an all-electronic computer.

ENIAC

Given the isolated nature of the work on Colossus, the research with the broadest impact in converting computing machines from mechanical designs to an all-electronic one was the ENIAC project. In 1944, John Mauchly, a professor at the University of Pennsylvania, set out to design a large-scale, general purpose, all-electronic computer. There would be no moving parts in the machine; all logic, computation, and data storage functions would use the switching function of triode circuits. Though all involved knew the difficulty of the task that lay before them, expectations were high that this new calculating machine would prove a major breakthrough in computing speed and capability.

Mauchly received his PhD in physics from Johns Hopkins University in 1937, after which he began teaching at a small university outside Philadelphia. He always had great interest in computing machines, and in 1940, while attending a conference on computational sciences, Mauchly met a young professor from Iowa State College (now known as Iowa State University) named John Atanasoff. Atanasoff told Mauchly that he had spent the last three years working with a student to develop a new type of computing machine. He invited Mauchly to visit his laboratory, an invitation that Mauchly accepted.

Mauchly traveled to Iowa and stayed at Atanasoff's home, where Atanasoff openly shared his thoughts, showing Mauchly all his detailed designs and explaining the rationale behind each decision. Mauchly spent four days carefully examining Atanasoff's machine, which was a small-scale machine employing only 280 vacuum triodes, a quantity restricted mainly by Atanasoff's extremely limited research funding from his university's general fund. Mauchly realized

John Atanasoff. Special Collections Department/ Iowa State University Library

that there were two crucial innovations in the architecture of the machine that put it ahead of other known computing machines. The first was that all data were represented in "binary" form, that is, with either a "1" or a "0." The second innovation was that all of the binary logic and computing functions were carried out via electronic switching mechanisms that used vacuum triode circuits. This principle of operation was based on Boolean algebra, an obscure branch of mathematics invented in 1854 by a British mathematician named George Boole. Instead of the familiar decimal numbering system, Boolean algebra was based on a binary numbering system that unified both mathematical and logic operations, using just 0s and 1s. When Boole first published his work, few considered it to be of any real utility, but Boole boldly predicted that his new mathematics would make an enormous impact on mankind. Almost a century later, he was proven correct. (Note: Colossus predated Atanasoff's machine in using vacuum triode circuits to process binary data based on Boolean algebra. But there was no way that Atanasoff could have known that at the time.)

Like many scientists and engineers of the time, in 1942, Atanasoff was drafted into the Navy. He was assigned to work on sonar technology, so he was unable to continue research on his computing machine. As he prepared to leave Iowa to report for duty, he asked his college's administration to assist him in filing a patent application for his novel design. However, to Atanasoff's later chagrin, the institution never followed through.

After Mauchly returned from his visit with Atanasoff, he felt much more confident in his conceived approach to design and build an all-electronic computing machine. In 1942, right around when Atanasoff was drafted, Mauchly was hired by the University of Pennsylvania, where he joined the Moore School

John Mauchly. Courtesy of John Mauchly Papers, Rare Book and Manuscript Library, University of Pennsylvania Libraries

of Engineering as a faculty member. While there, he met Presper Eckert, a graduate student in the electrical engineering department. Eckert was an outstanding young engineer, well-versed in both systems and circuit design, and with a bent toward experimentation. He seemed the perfect match for Mauchly as a working companion. It didn't hurt that both Mauchly and Eckert harbored grand ambitions of catapulting electronic computing technology to a new level and gaining fame and fortune for themselves.

Over the course of the next year, the two men designed a large-scale, completely electronic computing machine. In order to obtain funding support to build it, Mauchly reached out to the biggest funding source he could conceive of: the U.S. War Department.

Only one hour away from the University of Pennsylvania sat the Aberdeen Proving Ground, a branch of the U.S. Army Ballistic Research Laboratory. One of the duties of the Proving Ground was to calculate and validate data tables required for newly developed artillery. These tables presented pre-calculated specifications so as to provide men in the field with ready references of coordinates for aiming projectiles at targets under different types of terrain or weather conditions, including high wind.

When Mauchly visited the Proving Ground in 1943, there were more than 300 varieties of newly developed artillery and projectiles awaiting the results of firing calculations. Although the Proving Ground had more than 200 full-time staff (known as "computers") that specialized in performing such calculations with state-of-the-art electromechanical machines, they still could not finish that volume of work in a timely manner, and the pressure upon them mounted day by day.

Lieutenant Herman Goldstine was the technical officer at the Proving Ground responsible for liaison with universities. Goldstine himself had a PhD in mathematics from the University of Chicago, and he recognized both the predicament in which his charges found themselves and the potential of Mauchly's design. After evaluating the proposal carefully, Lt. Goldstine decided that it was worth a try, and he managed to wrest $500,000 from the U.S. Army Ballistics Laboratory budget to fund Mauchly's project. At the time, Goldstine hoped only that this high-performance machine would be able to speed up the calculations urgently needed to clear the backlog at the Proving Ground and assist the war effort. As it happened, this act had far greater reaching effects, and it would open the path to ushering in modern computer technology.

The project was launched in 1944, and the computing machine was named the Electronic Numerical Integrator And Computer—ENIAC. The word "computer" as referring to a physical construct, rather than a person, was used for the first time in the name of the machine.

Over the next two years, Mauchly, Eckert, and all of their associates became totally immersed in the intense effort to construct ENIAC. During that period, Mauchly made several special trips to visit John Atanasoff, who was stationed at the Naval Ordnance Laboratory in Washington, DC, to ask many technical questions regarding the design of Atanasoff's original binary computing machine. However, Mauchly never mentioned the ENIAC project or disclosed to Atanasoff the true nature of his work. In fact, Mauchly eventually patented many aspects of the computer—including some of those that he'd seen and learned about in Iowa—without ever crediting Atanasoff for his contributions.

ENIAC was not completed until 1946, at which point, the war was over. Thus, ENIAC never did play a role in the war effort itself. Nonetheless, it marked an important milestone in the development of electronic computing. ENIAC was, at it had been designed to be, all-electronic. In operation, it could perform up to 357 multiplications every second or up to 5,000 additions and subtractions, making it well over a thousand times faster than the Harvard Mark I. Additionally, its operating instructions could be reprogrammed to perform many different tasks. Of course, all this power came at a price. The construction of ENIAC required 17,468 vacuum triodes and more than 5 million welded joints. In operation, the machine consumed 160 kilowatts of power, more than Edison's entire groundbreaking Pearl Street power plant could have produced, and its weight topped 60,000 pounds.

Although ENIAC successfully demonstrated the feasibility of an electronic computer, it did suffer from many serious shortcomings. First, the computer's memory capacity was woefully inadequate, which stunted its potential to perform many important functions. This problem was not easy to address without the development of a new memory technology. Secondly, ENIAC's operating software was programmed externally using complicated cable connections and

ENIAC filled an entire room, center person was Eckert, with Mauchly to the right. Courtesy of John Mauchly Papers, Rare Book and Manuscript Library, University of Pennsylvania Libraries

switch position settings. Though software reprogramming was possible, any such work was both extremely slow and highly error-prone. Finally, ENIAC relied on too many vacuum triodes. Although the average lifetime of vacuum triodes during that time was around 3,000 hours, ENIAC used almost 18,000 of them. Statistically, this meant the average time to failure before a vacuum triode blew and the system shut down was no more than a few seconds, making the machine unreliable almost to the point of uselessness. In fact, the technical weaknesses of ENIAC were so clear during its development phase that by 1945, before the project was even completed, Mauchly and his team had already begun to design an improved, successor machine, called the Electronic Discrete Variable Automatic Computer, or EDVAC.

Despite its shortcomings, ENIAC was a historical achievement. By sheer brute force, it proved that creating a fast, general purpose, all-electronic computer was feasible. Additionally, its potential capability was scalable, limited only by the number of vacuum triodes that could be used in the machine. Experts began to salivate at the potential performance of computers implemented with millions or even billions of vacuum triodes. Still few dared to dream of a future that such powerful machines could be technically and economically feasible only a quarter of a century away.

FOUNDATION OF COMPUTER ARCHITECTURE

The intention of the U.S. military in developing an all-electronic computer was to solve a specific, short-term technical problem. The decision was made at a low level, and the total funding was commensurately modest. The project was very different from the top-down, strategic, massive investment made in the development of radar or the atomic bomb; however, its impact was equally significant.

During the early days of ENIAC's construction, Lt. Goldstine encountered by chance at the Aberdeen train station the eminent mathematician, Professor John von Neumann of the Institute for Advanced Study at nearby Princeton University. Goldstine began lamenting the unforeseen challenges facing ENIAC to von Neumann—but, unbeknownst to Goldstine, von Neumann was also employed by the War Department, working on the top-secret Manhattan Project. Professor von Neumann had a broad range of intellectual interests. In addition to applying advanced mathematics to fields such as quantum mechanics, economic modeling, and Game Theory, he was deeply interested in brain intelligence. So the ENIAC project aroused an interest in him.

John von Neumann. Science Source

Goldstine invited the professor to consult for the computer design team. After studying the overall design of ENIAC, it occurred to von Neumann that the architecture of the machine could be improved significantly by adding a central processing unit, or CPU. Furthermore, von Neumann became convinced that the clunky software programming problem could be solved by building the computer its own internal memory bank and storing the programs and the data at the same address space. This "stored program" approach was a revolutionary concept in computer design, and it vastly improved the overall functionality and flexibility of the computer.

Lessons gleaned from the shortfalls of ENIAC proved extremely valuable in designing subsequent generations of computers. For instance, although Mauchly had learned of the superiority of using the binary computing approach when he first visited Atanasoff, he inexplicably chose to implement a decimal system in ENIAC. Further study later made clear that using a binary design could significantly reduce system complexity. Starting from EDVAC, almost all computer designs were based on the binary system and the von Neumann architecture with a CPU, including the embedded stored program concept. John von Neumann himself played a big role in the design of the EDVAC, which inevitably led to significant personal conflicts with Mauchly and Eckert. Fortunately it did not derail the rapid technical advancement in computer development.

As EDVAC neared completion, a new problem arose: how best to disseminate all this knowledge about electronic computer design and spur the growth of a new computer industry? The Dean of the Moore School of Engineering at the University of Pennsylvania decided to organize a lecture series in the summer of 1946. He invited thirty-seven prominent computer scientists and engineers from various institutions to attend an eight-week "Moore Lecture" series. All contributors, including von Neumann, Mauchly, Eckert, and Goldstine, made detailed presentations. Concepts such as the stored program were described in full detail. The impact of the Moore Lecture series could not be underestimated. Most of the attendees later played key roles in the development of computer technology and the creation of the computer industry.

With the successful completion of the ENIAC and EDVAC projects, it appeared the all-electronic computer was ready to be introduced to the commercial world. Mauchly and Eckert, being entrepreneurs themselves, felt it was time to move on and make their fortunes using their knowledge. They applied for and were later granted the fundamental patent for electronic computers based on the ENIAC work, and with that patent, they claimed to be the inventors of the digital computer. Together, they formed a company in 1946 in Philadelphia to commercialize the computer. The first revenue for their company came from a contract with the Census Bureau to build a special computer for processing census data. The acronym for this computer was UNIVAC, for UNIVersal Automatic Computer, which they also adopted as the name for their new company.

UNIVAC pioneered the start of a string of new computer companies in the early 1950s. As business grew, a legal battle hinging on patent rights became inevitable. History always repeats itself, and as with the telegraph, the telephone, the television, and radio, ownership and the mantle of creation of the computer began to be a disputed topic. Other companies challenged Mauchly and Eckert's ENIAC patents in court, and finally, in the mid-1970s, their foundational patent was overturned. The credit for the invention of the digital computer was ceded to the man who rightfully deserved it: John Atanasoff.

Though Atanasoff at last was afforded the accolades he deserved, he never reaped a commensurate financial windfall. Since Atanasoff was the inventor of the computer but neither he nor Iowa State College had ever followed through in applying for a patent, the courts ruled that the patent rights would be assigned to the public domain. This ruling allowed any individuals or companies to develop computer products without having to worry about basic patent infringement, clearing the path for the rapid growth of the computer industry.

FRAMEWORK FOR THE FUTURE

The U.S. Army's investment in the ENIAC and EDVAC projects established the foundation necessary to create the modern electronic computer. Because of the high cost and high level of risk involved—especially given the unclear commercial market potential for such a novel product—private enterprise would likely have been reluctant to invest in such a project. Even the most laissez-faire economist and businessman must take note that sometimes it takes government involvement to push past such catch-22s and help create something truly revolutionary. It was true of the telegraph, it was true of radar, and it was true of the computer—and later, it would be true of the Internet. It will surely be true of some other game-changing innovation in the future as well.

It would be impossible to discuss computers without mentioning International Business Machines—IBM—a company that can actually trace its roots to mechanical computers. In 1888, Herman Hollerith, the American pioneer for mechanical computing machines, founded the Tabulating Machine Company. This predecessor company to IBM won a contract from the U.S. Census Bureau to build punch card equipment to help tabulate results from the 1890 Census. IBM continued to gain computer expertise in the 1930s through collaboration with Harvard on the Mark I computer.

At the height of the Cold War, IBM worked closely with MIT and the U.S. Air Force to develop the most advanced and complex computer system, code-named SAGE, which was a large-scale air defense system, linking a large amount of information from multiple radar systems across the country to a centralized command and control center. This experience allowed IBM to be at the cutting edge of computer hardware and systems software technology.

IBM owned a unique culture that put special emphasis on delivering total solutions to satisfy customer needs. The company recognized early on that the applications for powerful computers lay not just in complex scientific and engineering calculations, but also in the more mundane processing and analysis of massive amount of repetitive information. As the amount of information in the business world swelled exponentially in the postwar era, the market potential for computers grew correspondingly. Starting from the early 1950s and for several decades later, IBM was the indisputable leader in the global computer industry. It even played a key role in starting the personal computer revolution. Along the way IBM also pioneered many foundational computer hardware and software technologies, including the magnetic hard disks for mass data storage, which played a key role in today's information age.

III

AGE OF SOLID-STATE ELECTRONICS

12

THE SEMICONDUCTOR

BELL LABS

In the first half of the twentieth century, nearly every major electronic invention, including radio and computer, was enabled by the unique signal amplification and switching functions of the vacuum triode. However, the triode's inherent limitations—high cost, short operating life, easy breakability, and cumulative size and power consumption—were severe constraints to the continuous scaling up and improvement of many of these innovative creations. The ENIAC computer, impressive achievement though it was, presented clear reminders of this bottleneck.

To overcome these obstacles, the burgeoning computer and electronic industry required a new, fundamental breakthrough: a signal amplifying and switching device that could perform significantly better than the vacuum triode. The quest for this revolutionary device eventually ushered the electronic world into its modern age. The cradle of this revolution was a unique and remarkable research organization located in the peaceful rolling hills in central New Jersey: Bell Laboratories.

When Ted Vail returned to AT&T in 1907, one of his key strategies was to solidify the company's position as the absolute leader in communications technology. He was determined to never again allow his company to grow beholden to the intellectual property of other inventors and firms. To that end, Vail founded a laboratory in New York City under the oversight of AT&T's equipment manufacturing subsidy, Western Electric. Vail staffed this technology division with the brightest talent that he could recruit across all fields of engineering. By design, the breadth of research projects undertaken at the laboratory was vast, ranging from physics, chemistry, and materials sciences to electrical engineering, mathematics, information sciences, and network systems. Though it may not have been clear to those who lacked Vail's extraordinary vision, linking all of these seemingly diverse disciplines was part of a focused, singular concept to improve and expand the frontiers of communications technology.

With its deep understanding of both technological and marketplace needs, this AT&T team unearthed De Forest's vacuum triode technology from the brink of obscurity in 1911, identifying it as the solution to breaking through the bottleneck then throttling long distance telephony. After purchasing rights to the device, the team showed its unique value by quickly refining the triode and successfully applying it to the trans–North American long distance line. By 1925, the Western Electric technology department was officially renamed Bell Laboratories. Ever successful and ever growing, Bell Labs quickly grew too big for its surroundings in New York city, and as the organization expanded, the department pulled up stakes and moved to its current, more expansive locations in the rolling countryside of central New Jersey.

Bell Labs at Murray Hills, New Jersey. Reprinted with permission of Alcatel-Lucent USA Inc.

The funding model for research at Bell Labs was unique. Because AT&T continued to operate as a government-sanctioned monopoly, the fees it charged for telephone service were subject to periodic review by a publicly appointed commission. The commission would examine AT&T's proposed operating budget and, after scrutinizing the company's financials, allow AT&T to establish rates, permitting it to cover its approved costs and earn a reasonable but modest profit on top. Thus, Bell Labs' research funding was essentially shielded from any competitive market forces. In addition, research costs comprised only a small percentage of the total business expenditures of this behemoth organization, and as those costs were baked into total operating needs by the public auditors, AT&T's research budget remained largely and enviably unaffected by financiers' demands for short-term operating profit. Bell Labs thus enjoyed the unique stability necessary to pursue multiple long-term and high-risk—but potentially high pay off—research projects. Since nearly all of Bell Labs' research projects had clear potential applications in communications systems and strong relevance to AT&T's core business, the company was also far more successful in transitioning the results from their in-house research into impactful products and services than most other research organizations, most of which had abstract and diffused goals and lacked a viable market pull. These factors served to make Bell Labs much more attractive to prospective recruits: what top-flight technical talent wouldn't be interested in working at the world's best-funded and most prestigious lab, and with the best chance to create new technologies that had a lasting impact upon the society?

KELLY'S FORESIGHT

In 1936, Mervin Kelly was appointed Bell Labs' Director of Research. He had been with the organization since 1918 after getting his PhD in physics from the University of Chicago. During his eighteen-year tenure leading up to this promotion, he had, among other roles, served as a key researcher in the very successful vacuum triode development group. Over a period of twenty years, this team managed to increase the average lifetime of triodes from 800 operating hours to 80,000, a hundred-fold improvement. But by the mid-1930s, almost every aspect of the triode's performance had been optimized, and Kelly, who knew more about triode technology than anyone, could see all too clearly that the technology had progressed as far as it could. For AT&T to be able to develop the envisioned future generations of communications capabilities—for

Mervin Kelly. Courtesy of AT&T Archives and History Center

instance, a fast, automated, large-scale central telephone switch, or a reliable, transatlantic undersea telephone cable—it would have to develop a next-generation device, one that could replace the vacuum triode. But what kind of new technology would it be? The answer was unclear, but the deeply insightful Kelly had formed a vision based on a curious discovery that had been made some six decades earlier. In 1874, German scientist Karl Ferdinand Braun accidentally discovered that a metal-galena point-contact device could exhibit a current-rectifying property, similar to Fleming's vacuum diode. By adding a grid electrode to the vacuum diode, De Forest turned the diode into a vacuum triode. By analogy, would it be possible to somehow insert an electrode between the metal and the galena to create an equivalent solid state triode?

Kelly's exceptional intuition and vast experience convinced him that it should be possible to develop such a solid device that performed like a vacuum triode. It was an excellent notion, but Kelly also knew that the technical challenges would be enormous and the risks of failure would be equally high. The key question for him was how would Bell Labs go about actually developing such a revolutionary device? Kelly's answer was simple: build a first-rate technical team with a clear goal, provide it with strong management and financial support over the long term, and let the team find the solution on its own. He also had the instinct that the key to solving the problems was in developing new understandings in the fundamental physics and material sciences.

THE UNPREDICTABLE SEMICONDUCTOR

Karl Ferdinand Braun's accidental discovery of the metal-galena point-contact rectifier marked the beginning of solid-state electronics. Unfortunately at the time, the phenomenon was unstable, unpredictable, and difficult to reproduce. Not all galena succeeded in allowing unidirectional current flow, and even for samples that did exhibit rectification properties, the phenomenon might occur only on a few spots across the entire sample surface. During use, one had to employ a fine brass wire splayed at the end like a cat's whisker (hence its common name of "cat whisker rectifier") to probe the sample surface and locate spots that exhibited the right rectifying properties—quite a frustrating process. There was no practical use for the cat-whisker rectifier until in 1900 when Indian inventor Jagadish Bose discovered that he could use such rectifiers to detect wireless signals with excellent sensitivity. Later, as radio broadcasting became popular, the cat-whisker rectifier would be widely used by radio hobbyists to build simple—if at times exasperating—crystal radio receivers, which was a cheaper alternative to vacuum diodes.

As crystal radios gained popularity, people began to search for a better material to replace galena, which was a natural crystal of lead-sulfide (PbS). Soon some scientists discovered that there were many other substances—including the elements silicon (Si) and germanium (Ge), both in column IV of the periodic table (group 14, by the new IUPAC numbering system). Because these materials were in some instances electrically conductive while in other instances they were not, they came to be collectively known as *semiconductors*.

Karl Ferdinand Braun. Science Source

As more and more experiments were performed, it became clear that the conducting properties of semiconductors were greatly affected by even minute amounts of contaminants, or minor structural defects in the material, which explained why the semiconductors themselves behaved so unpredictably. This complexity troubled researchers, and scientists realized that they would have to learn more about the fundamental physics of these materials at the atomic level in order to better understand them and gain the capacity to predict and control their behavior.

It wasn't until the late 1920s that scientists began to apply the newly developed principles of quantum mechanics to interpret the various properties of condensed matter, or solids. This study of *solid-state physics* was aided by the development of a new measurement technique employing X-ray diffraction, which could precisely measure the exact atomic structures and dimensions of different solids. Applying this newly acquired information, a German scientist named Paul Drude developed a theoretical model to calculate and successfully predict the properties of various metals. In 1931, Britain's Alan H. Wilson commenced his own, related work, concentrating not on metals, but on semiconductors. In his studies, Wilson used quantum mechanical principles to predict that the energy distribution of electrons within a semiconductor lay across a series of bands, with forbidden "energy gaps" between each. Free-flowing electrons that possessed energy levels placing them within these identified energy gaps could not exist in pure and perfectly structured semiconductors. The magnitude of these energy gaps varied amongst different semiconductors, and they, along with impurities in the material, determined the semiconductors' basic

Alan H. Wilson. Courtesy of Master and Fellows of Trinity College, Cambridge

Walter Schottky. Siemens Corporate Archives, Munich

properties, such as their conductivity. Wilson's seminal work showed the power and impact of quantum mechanics, and after its publication, theoretical research into semiconductors blossomed into an important branch of solid-state physics.

Building on advances made by his peers to further the understanding of semiconducting materials, German scientist Walter Schottky began to apply the theories of solid-state physics to explain the behavior of the point-contact rectifiers. Schottky was a university professor as well as a senior researcher at Siemens. In 1938, he published a seminal paper entitled "Metal-Semiconductor Junction Rectification" that successfully explained the basic physics of the interface between metals and semiconductors, including the cat-whisker galena rectifiers. According to Schottky, the rectification phenomenon was caused by an energy barrier that was formed at the interface of the metal and semiconductor, which inhibited the flow of electrons in one direction, but not the other. This barrier was later named Schottky Barrier, and the metal-semiconductor contact rectifier itself came to be known as the Schottky diode.

Schottky's paper was an epochal piece of work in semiconductor technology: it marked the first time the basic principles of physics were successfully employed to explain the observed functional behavior of a semiconductor device. Schottky's theory could also explain why some contact points on a galena surface were rectifying while others weren't. His work was nothing short of groundbreaking; however, his timing was poor. Almost immediately after the publication of his paper, World War II erupted, and all international academic exchange ground to a halt. This left Schottky's astounding achievement temporarily overlooked as the globe turned its focus toward conflict. As the Nazis rose to power in Germany, many Jews—including Jewish scientists and academics—

fled Europe and found permanent homes in America. Many of the scientists were recruited into American universities, greatly strengthening the depth and breadth of the research capabilities. This movement also led to a subsequent butterfly effect: the creation of a still-broader talent pool in solid-state research, as American students received top-notch teaching and training from experts in the field hailing from all four corners of the earth.

13

THE BIRTH OF THE TRANSISTOR

THE FLAMBOYANT GENIUS

In 1936, even before Schottky had published his seminal paper on the physics of metal-semiconductor contact, Mervin Kelly was able to intuit that it might be possible to develop a solid-state triode based on the point-contact rectifier. At the outset, Kelly did not have the luxury of a clear, specific approach to pursue his goal. He also knew full well that the technological challenges that lay ahead would be enormous. But he was convinced that research in solid-state triodes held great potential, and if successful, the impact would be huge. So he pressed on.

Kelly fully understood that assembling a technical team that was both first-rate and multidisciplinary would be key to his success. The ideal team would consist of outstanding electrical engineers, semiconductor materials experts, chemists, and experimental physicists, to say nothing of the crucial need for a top-notch theoretical solid-state physicist who could see the big picture and serve as team leader. While Bell Labs' high esteem and salary offerings ensured that they were able to recruit the best—that very year, a Bell Labs' employee, Clinton Davisson, won the Nobel Prize in Physics for his research on electron diffraction—there was unfortunately no potential candidate at the company to lead the solid-state triode project. Therefore, Kelly looked outside the company, where he found a brash, young physicist from MIT named William Shockley.

William Shockley was born in England in 1910. Despite his birthplace, he came from among the oldest American stock: his ancestors had come over on the Mayflower. Shockley's father was a mining engineer who graduated from MIT, and his mother was among the first female graduates of Stanford Uni-

William Shockley. Emilio Segrè Visual Archives/American Institute of Physics/Science Source

versity, where she studied mathematics and art. Despite the similarities in their backgrounds, there was a twenty-five-year age difference between them, which may have led the Shockley family to be viewed with some public suspicion. As a result, they did not do much socializing with other people. In addition, William was the only child of the couple, which may have propelled him even further to exhibit a propensity toward introversion. While the young Shockley demonstrated his high intelligence early on in life, he was also rebellious and aloof.

After returning to America from England, the Shockley family settled in Palo Alto, California, near Mrs. Shockley's alma mater. William Shockley later attended and graduated from the prestigious California Institute of Technology (Cal Tech), building a strong foundation in physics and mathematics before culminating his studies with a PhD in solid-state physics at MIT. As something of an intellectual wizard who lacked social finesse, Shockley could at times be grandiose, even insufferably so. He held himself in high regard and was extremely ambitious; he was determined to do something big in life. He liked fast cars, guns, and playing magic tricks; he also liked facing challenges, including climbing difficult rock faces. Shockley would come to claim that the greatest joy in life was climbing up a sheer cliff bare-handed on a moonlit night. Even today, there is a climbing route in the Shawangunk Mountains in the Hudson Valley region of New York named after him.

In 1937, Kelly persuaded Shockley, who was then just twenty-seven years old, to join Bell Labs, envisioning that this brilliant young man with a top academic pedigree might be the one to bring Kelly's vision to reality. On Shockley's first day on the job, Kelly took him aside and encouraged the new recruit to use his great intellect and ability to develop the solid-state triode. The triode, Kelly explained, would replace the vacuum tube, and then that technology could be applied to build an all-electronic central telephone switch, bringing enormous benefit to both the company and society as a whole. Never mind that nobody knew whether or not such a device was even possible. Kelly showed great confidence in Shockley, assigning him to begin work with the Nobel Prize winner Davisson in the vacuum tube research group, but also giving him the freedom to get involved in any of Bell Labs' projects as he saw fit. Kelly encouraged Shockley to look for his own opportunities via which to succeed, and this was a conversation that Shockley was to remember throughout his life.

CONCEPTUALIZING A SOLID-STATE TRIODE

Shockley threw himself into his work with a passion, scrutinizing every project Bell Labs was then currently undertaking and reading all the papers about semiconductors he could get his hands on—including Walter Schottky's "Metal-Semiconductor Junction Rectification." Of all the research efforts, the one that most interested him was the cuprous oxide (Cu_2O) semiconductor project led by Bell Labs' senior researcher Walter Brattain. Shockley sensed that by combining Schottky's concept with Brattain's experimental work on cuprous oxide, it might be possible to create a solid-state triode that could indeed amplify signals, just like vacuum triodes did. Shockley spent several months consolidating his ideas and designing a device. He eventually decided to use a small rectangular piece of semiconducting cuprous oxide, placing conducting electrodes at both of the long ends. He then added a third metal electrode above one of the long surfaces of the sample, but separated the metal electrode and the semiconductor surface with a very thin layer of insulating material. Based on Schottky's theory, Shockley predicted that the resistance of the cuprous oxide slab between the two end electrodes could be modulated by a voltage applied to the third "gate" electrode, functioning just like a "grid" electrode would in a vacuum-tube triode. He was certain that the *electric field* from the third electrode would effectively penetrate into the semiconductor material underneath, modulating the resistance of the sample and the current flow between the two electrodes. (This design was in fact conceptually analogous to Edison's carbon granule microphone, with the membrane for acoustic signal input replaced by the gate electrode, and the granular carbon resistor replaced with a piece of semiconductor.)

Shockley's design was simple and elegant. His experimental attempts, however, failed entirely. This setback frustrated him. He was not accustomed to

Walter Brattain. Reprinted with permission of Alcatel-Lucent USA Inc.

being wrong, and he had been very confident about the theoretical soundness of his *"field-effect"* solid-state triode. Though he was willing to admit that his experiment was a failure, he was adamant that the flaw was not in his design. At last recognizing that experimentation was not as strong a suit to him as theory, he approached for help the man whose work had first led him toward cuprous oxide: Walter Brattain.

Born in Xiamen, China, in 1902 where his father taught mathematics and physics, Walter Brattain returned to America with his family when he was one year old. He grew up on a ranch in Washington State not far from the Canadian border, idolizing and living the cowboy culture: tough, independent, and straightforward. Influenced by his father, Brattain developed a deep interest in physics from a very young age. Often, Brattain would play with his crystal radio set after a day spent riding horseback and rounding up herds of cattle. When he grew up, he went to college, and after receiving his PhD in physics in 1929 from the University of Minnesota, Brattain joined Bell Labs.

By the late 1930s, Brattain had developed a reputation as one of the best experimentalists in the company. Dexterous and creative, he also possessed a wealth of experience and a keen intuition, so it was no accident that Shockley approached him for help in refining his solid-state semiconductor triode. Though Brattain felt sure that Shockley's cuprous oxide design would not work, he nonetheless patiently went along with what the brilliant, flamboyant novitiate wanted, fabricating a new three-terminal device according to Shockley's specifications. As they began to test the new invention, Shockley watched intently as Brattain connected the wires and then ever-so-slowly turned up the voltage to the device's third gate electrode. However, the needle on the meter measuring the resistance of the cuprous oxide slab did not make the slightest movement. Its electrical resistance was completely unaffected by the increased voltage applied to the gate electrode.

Though this second failure came as no surprise to Brattain, Shockley was deeply disappointed. Still, he learned a valuable lesson: there lies a huge gap between an idealistic theory and the complexity of reality. Semiconductor materials technology at the time was indeed primitive, and simply having a good device design on paper was not enough. The basic properties of semiconductors were affected by many factors, including material composition, external electrical fields, surface conditions, ambient temperature, traces of impurities in the material, and physical structural defects. Shockley and Brattain knew that some of these factors surely played a role in their failure, but they had no clue which parameters were the culprits, nor did they have the ability to control them—the technological problems were just too complex to overcome. Of course, they didn't have much time to figure out the matter: before the close of 1941, Japan bombed Pearl Harbor, and America found itself dragged into war.

FORGING A BETTER SEMICONDUCTOR

In 1940, the Tizard Commission brought the most valuable luggage, the new magnetron, to the United States, seeking joint development of radar technology. Starting from that point, the British and American technologists began to work closely together to further enhance radar performance and develop production capabilities. In the early days of their collaboration, the British-made radar receivers consistently outperformed American versions. When the U.S. engineers were finally able to carefully examine British radar receivers, they realized why: the British were not using vacuum tubes in their microwave receivers, but rather point-contact rectifiers, or Schottky Barrier diodes, made from purified silicon or germanium crystals. These rectifiers, when working properly, were far superior to vacuum tubes in terms of sensitivity and speed of response to microwave radar signals. From that point on, the United States began a massive effort coordinated across multiple organizations to improve the performance of silicon- and germanium-based point-contact rectifiers. Bell Labs was a part of this intensive effort.

When the United States entered World War II, the nation went into full-scale mobilization. Many Bell Labs' technologists were called into service, taking part in various projects developing critical military technologies. Shockley himself joined the Navy's antisubmarine warfare group. Not long after that, he was transferred to the War Department headquarters in Washington, DC, to develop a new branch of mathematical science known as Operations Research. The aim of this effort was to develop mathematical algorithms and models for use in analyzing and optimizing large-scale military operations and in improving logistics systems designs.

Shockley and many others had been called into service, but one of the few teams at Bell Labs that remained intact during the war was the one responsible for developing high-frequency radar and radio communications technology, where much emphasis was placed on improving the performance and manufacturing yield of point-contact rectifiers. This project group was managed by a senior materials engineer named Russell Ohl.

Ohl focused his team's efforts on systematically comparing the performance of point-contact rectifiers made from various types of semiconducting materials. In the process, he discovered that silicon and germanium consistently performed the best; they were far superior to rectifiers made of other semiconducting materials like lead sulfide (galena) or cuprous oxide. However, even among rectifiers made from silicon or germanium, levels of performance varied widely, even among rectifiers fabricated using identical methods from the same batch of material.

When sorting through the data, Ohl noticed that rectifier performance tended to correlate with the purity level of silicon or germanium used, and in general, the higher the purity, the better the performance. He decided to try to create the most pure silicon possible in his own lab. To that end, he purchased the highest grade silicon (>99.99% pure) commercially available; he then further purified the materials by melting them in a quartz tube at very high temperature, followed by long periods of baking under a continuously pumped high vacuum, with the hope to expel all volatile impurity atoms from the melt.

Ohl was a metallurgy expert, and he knew that silicon needed to shed heat very gradually during the cooling process in order to avoid any thermal shock that might lead to internal cracking as it recrystallized. Therefore, Ohl vigilantly controlled the temperature of the silicon to allow it to cool to room temperature over an unusually long period of time. Afterward, he cut and polished the super-purified silicon into standard samples for testing. However, to his surprise, some of these carefully prepared samples behaved strangely when he tested their conductivity. Often, it was impossible to obtain a stable reading from them, and even more mysteriously, he found that in some instances, a significant electrical current *could* be detected to flow across the sample when light was shone on them. What did these results mean? Ohl had no idea, so he decided to reach out to the director of the Research Department at Bell Labs: Mervin Kelly.

Kelly realized readily that Ohl's observation could be extremely valuable; perhaps it was an important clue in solving the semiconductor puzzle. What exactly it meant, however, he wasn't sure, so Kelly brought together all of the Research Department's top technical experts to witness Ohl present and demonstrate his findings. By that time, Shockley had already left for the Navy, but Walter Brattain—despite having already received notice to report to the Naval Research Lab to participate in developing magnetic sensing devices for homing torpedoes—was still present.

After watching Ohl's demonstration, Brattain pondered for a moment, then postulated that perhaps this unusual behavior was the result of an energy barrier inside the silicon sample that inhibited electron flow, a barrier much like the one between semiconductor and metal that Walter Schottky had discovered in 1938. He also suggested that when light struck the sample, it generated electrons which were free to flow, altering the magnitude of this barrier and inducing the sample to produce an electrical current. As to the possible cause and physical nature of this barrier, Brattain proposed that it could be related to the spatial distribution of the few atoms of impurities that did remain in the silicon sample.

Ohl was impressed by the suggestions—they were clear, rational, consistent, and pointed to a path for additional research. He agreed to continue collaborating with the staff of the Research Department with the hope that they might solve this mystery together. Mervin Kelly, too, was impressed—seeing the power of the combined intellect of his scientists working as one, he felt a tremendous sense of pride about his people and his organization. The team was on the right track!

DISCOVERY OF THE P-N JUNCTION

Over time, the joint Bell Labs team was able to learn much about the properties of semiconductors. When silicon is completely free of impurities and is structurally perfect, it conducts almost no electricity. Due to its energy gap—that unique property of semiconductors first suggested by Alan Wilson in 1931—it turns out that silicon contains very few electrons that can flow freely and contribute to conductivity at ambient temperature. However, if some small amounts of impurities are added to silicon—even only a few rogue atoms amongst millions—the conductive properties of the resulting "doped" material will greatly change.

The electrical conductivity of a pure silicon sample is directly proportional to the quantity and type of impurities added. For example, if an element like phosphorus from group V of the periodic table is added to silicon, each atom will contribute one free-flowing, negatively-charged electron to the silicon host. Similarly, if elements from group III such as boron are added as impurities, these atoms can "absorb" electrons in the silicon host and create empty, positively-charged spaces—known as "holes"—that function effectively as mobile particles of positive charge. Thus, in semiconductors, there are two different types of charged particles flowing simultaneously: negatively-charged electrons and positively-charged

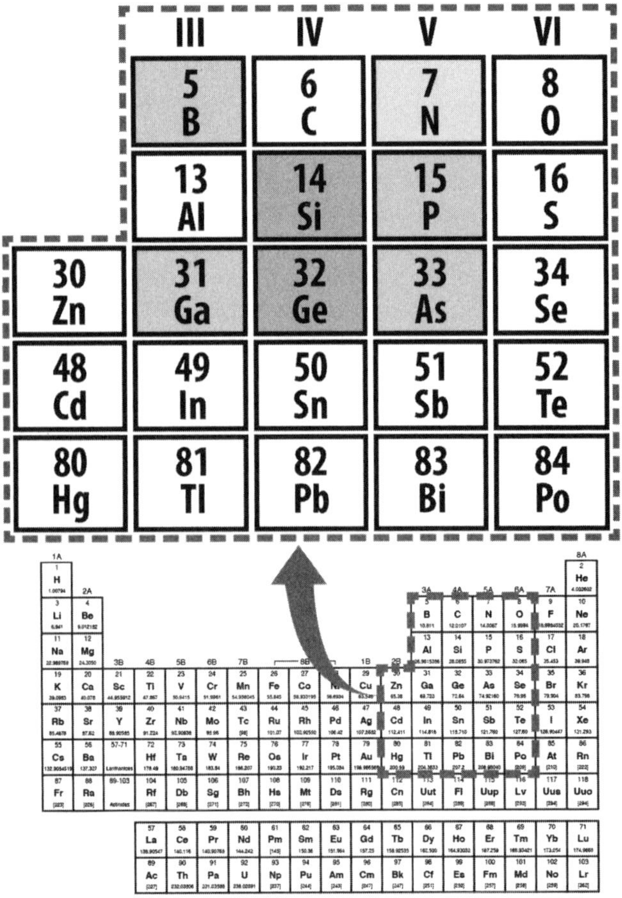

Periodic table, with enlarged portion illustrating group IV semiconductors silicon and germanium, group V n-type impurities phosphorus and arsenic, and group III p-type impurities boron and galium. Derek Cheung

holes. Their quantity is precisely determined by the type and amount of impurity atoms present within the silicon host. Armed with this knowledge, it became clear that very sophisticated materials technology would have to be developed in order to reliably and precisely control the properties of semiconductors.

Bell Labs' research staff called the semiconductors with excess electrons n-type (for negative) and the semiconductors with excess holes p-type (for positive). Ohl's team and the staff of Kelly's Research Department put forth an enormous collaborative effort to understand the roles of impurities in influencing the behavior of semiconductors and to develop techniques for controlling their presence.

Readily intuiting the value and importance of this work, Kelly swore everyone to total secrecy regarding Ohl's discovery, cautioning his employees to keep their findings secret from even the federal government. He was thereafter secure that his men could continue to accumulate knowledge free from the pressure of competing teams.

Thanks to the advances Bell Labs had made, the odd properties of the sample that first sparked Ohl to seek outside help could be understood. It turned out that part of Ohl's sample was p-type and the other part was n-type. This was why it had exhibited such strange properties. The residual impurity atoms of phosphorus and boron were initially uniformly distributed in the sample. However, during the slow cooling-down period from melt to solid, the impurity atoms segregated themselves due to metallurgical principles: the phosphorus atoms were primarily on one side, and the boron atoms were primarily on the other. It was Ohl's meticulousness that had allowed this odd—and valuable—behavior to occur.

Additional testing of the sample found that there existed a self-generated electrical field in the sample at the interface, or junction, between the p- and n- regions. This electric field balanced the density of electrons and holes on the two sides, and it was this self-generated electric field that formed the electron barrier that Brattain had first postulated. When voltage was applied to reduce the barrier, electrons and holes would spill over the barrier, causing a current to flow across the junction. When an opposing voltage was applied to increase the barrier height, no current would flow. Thus, the p-n junction allowed current to flow in only one direction—in other words, it successfully exhibited the property of current rectification in a single piece of silicon.

As for the curious effect of light inducing current, the team discovered that the silicon absorbed some of the light energy which created excess pairs of electrons and holes. When these free carriers diffused to the edge of the p-n junction, they were immediately separated by the built-in electric field, creating an external electrical current. This was the source of the photoinduced current that Ohl originally observed, and far from a purely academic curiosity, this technology would later serve as the basic physical principle behind photovoltaic solar cells and imaging sensors in electronic cameras.

ROADBLOCKS

Although Shockley was not physically present at Bell Labs during this period, he was elated by Ohl's discovery and followed the team's progress every step of the way. Many budding ideas raced about in his brain as he considered how to make use of the p-n junction to design a solid-state triode, as he'd initially been hired to do.

In 1945, as the war was coming to an end, Shockley and many of his contemporaries returned to Bell Labs. In addition to their old colleagues, they were now joined by many new recruits—scientists and senior engineers who decided to sign on at Bell Labs after their wartime military assignments were over, an infusion of new talent that further strengthened the solid-state electronics research team. Among the new hires was a man fresh from the Navy's Weapons Research Lab named John Bardeen. Bardeen, like Shockley, was a prominent theoretical solid-state physicist. In fact, Shockley had always admired Bardeen's exceptional physical insights, and it was Shockley who persuaded Bardeen to join Bell Labs after the close of the war.

John Bardeen. Physics Today Collection/American Institute of Physics/Science Source

John Bardeen was born in Wisconsin in 1908. His father was the dean of the medical school at the University of Wisconsin. As a boy, John was a mathematical genius. He entered university at the age of fifteen, majoring in electrical engineering. Following graduation, he worked at an oil company for several years, where he specialized in analyzing petroleum-bearing geological structure data. Later, he decided to return to school to pursue advanced studies in physics. His PhD thesis was completed under the guidance of Professor Eugene Wigner at Princeton University, one of the most outstanding theoretical physicists of the twentieth century.

Though the two men were equally brilliant, Bardeen's personality was drastically different from Shockley's. He was a gentleman, modest, and unusually low-key. Apart from work, Bardeen's only hobby was to play a round of golf on the weekends—quite different from Shockley's passions of rock climbing, fast driving, and playing around with handguns. Thus, though they respected each other professionally, Bardeen and Shockley were highly dissimilar people and not personally close. Bardeen and Brattain, on the other hand, hit it off immediately; they even actively agreed to share an office with one another.

After the war, Kelly reorganized Bell Labs' research department, setting up a new solid-state electronics research group with Shockley as the coleader in charge of research activities. Bardeen, Brattain, and many other outstanding researchers became Shockley's nominal subordinates, even though they were effectively autonomous as senior researchers, and Shockley's role was largely hands-off.

Coming together, Bardeen and Brattain discussed how to best restart the solid-state triode research project that had been halted due to the war. The two men decided to start from Shockley's original "field-effect triode" concept proposed in 1939. Though the initial effort had not been successful even with Brattain's assistance, the pair, armed with better theoretical understanding of semiconductor physics and the availability of much improved semiconductor materials, felt that they still ought to revisit this initial idea. Ever a bit of an egotist, Shockley of course lent his full support to this proposal.

Using high-purity and structurally near-perfect germanium and silicon crystals—thanks to the advances made in servicing the needs of radar receivers during the war, there was no longer a need to use capricious cuprous oxide—Brattain quickly fabricated solid-state triode test devices. Even though the original model hadn't worked, Brattain no longer doubted that this approach held merit. But when Brattain and Bardeen began to run tests, the results were the same as they had been seven years prior: the device simply did not work. The electric field applied from the gate electrode had no effect whatsoever on the resistance of the germanium or silicon slab underneath, despite the improved quality of materials used and the advances made in the understanding of semiconductor physics. The two men repeated their experiment from beginning to end multiple times, but the outcome was always negative.

Heading back to the drawing board, the pair painstakingly analyzed their results. In time, Bardeen rightly theorized that perhaps the reason for the failure was that a sheet of electrons had become trapped and accumulated near the surface of the semiconductor under the gate electrode. If this were true, the electrical field from the gate electrode might not be able to penetrate through its shielding. Based on this line of thinking, Bardeen developed a new theory that explored the physics of electrons trapped near the surface of a semiconductor, and the two men shifted the focus of their research toward finding a way to reduce the effect of trapped surface electrons blocking an externally applied electric field.

In spite of Bardeen and Brattain's slow rate of progress, morale was high throughout the entire solid-state electronics group. It was rare for any research organization to be staffed with so many top-notch experts from multiple disciplines, and the intellectual content of exchanges among team members was consistently rich and inspiring. Everyone seemed to enjoy working with each other. Shockley, though aloof as ever, had a broad range of personal interests, so he often kept himself busy participating in activities outside the group. Of course, even though Shockley chose not to get directly involved with the project, it did not mean that he wasn't keenly interested in the solid-state triode. In fact, his mind would frequently wander to the p-n junction, and he often quietly searched for ways to utilize its properties for signal amplification.

THE GREAT BREAKTHROUGH

Based on Bardeen's theories, by the late autumn of 1947, Brattain had developed a practical technique to measure the quantity of electrons trapped at the surfaces of semiconductors. In one experiment, he discovered that a piece of germanium on which water vapor had condensed showed a much lower level of trapped electrons than expected. Was it possible that the mobile ions in the water droplets may have done something to the trapped electrons at the sample's surface? On November 17, in another experiment, Brattain and Bardeen immersed a piece of germanium in an electrolyte solution, hoping that the mobile ions within the solution would neutralize the effect of electrons trapped at the germanium's surface. To their amazement, when they applied voltage to the sample in solution, they found that—for the first time—the externally imposed electrical field could penetrate the semiconductor's surface charge shield and influence the electrical properties of the bulk material underneath. This result in essence proved the feasibility of Shockley's original concept of the field-effect triode.

Bardeen and Brattain were extremely excited about this breakthrough, and they went on to perform several more important experiments to gather additional data. Finally, they combined all the knowledge they had acquired to design and build a new test device. They selected a slab of high-purity n-type germanium crystal and converted a thin surface layer to p-type. A gold film was then applied to the back of the germanium slab to serve as the base electrode.

Next, Brattain crafted a piece of prism-like plastic, and attached a very thin gold film onto two contiguous sides of its perimeter. He carefully slit this metal film along the apex of the prism, barely separating the gold film via a narrow chasm just 50 micrometers (50 millionths of a meter) wide. Finally, Brattain used a spring to press the top of this prism onto the germanium semiconductor's surface, until the gold film on both sides of the prism made good electrical contact with the germanium underneath. Now, the test device had three electrodes—the two gold strips on the prism and the gold contact at the back of the germanium sample.

On December 16, 1947, Brattain sat at his lab bench and carefully connected the electrical wires to the test device. Bardeen, holding a notebook, stood behind Brattain, tensely watching the meter readings on the measurement instruments. First, Brattain applied a one-volt positive bias between the first gold contact (the emitter) and the back electrode (the base) of the germanium sample. He then applied a ten-volt negative bias through a resistor between the second gold contact (the collector) and the germanium back electrode. At this stage, the

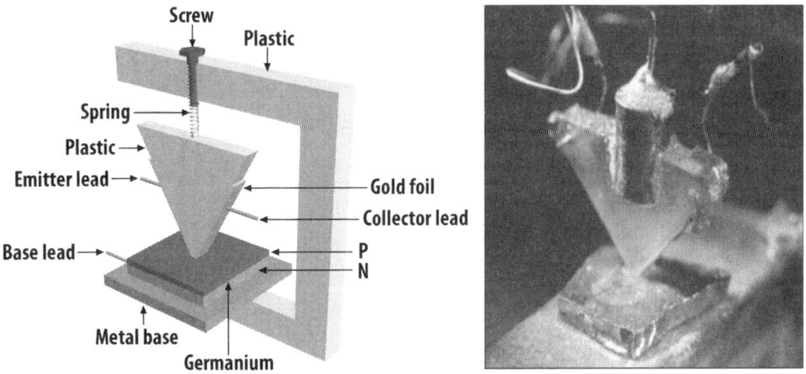

The first point-contact transistor by Brattain and Bardeen (right), and the schematics of the experiment (left). Reprinted with permission of Alcatel-Lucent USA Inc., schematic by Derek Cheung

device was behaving normally. Next, Brattain carefully connected a small oscillating voltage at 1,000 Hz in series to the emitter electrode. All of a sudden, the meter measuring the voltage across the carbon resistor at the collector electrode began to detect an amplified signal at 1,000 Hz. Bardeen was shocked. Could it be true? He checked again—it was! The long awaited moment had finally come: a solid-state amplifier was born! They looked at each other in disbelief with all kinds of emotions rushing through their minds.

After the excitement settled, Bardeen murmured, "We should call Shockley."

Brattain considered this. "Tomorrow," he replied.

That evening, Bardeen took his wife by the shoulders when he returned home and said to her, "Today, we have done something great!" Bardeen was by nature a low-key individual who never talked about his work at home, so she immediately knew that whatever Bardeen had accomplished that day, it was no ordinary matter. Brattain, too, told all his buddies in his carpool on the way home that he had done something extraordinary, something that might even change the world. He refused to elucidate on what it was, but that night, he couldn't sleep at all.

The next morning, the first thing that Bardeen and Brattain did was to call Shockley. Shockley took the news with mixed emotions. He was excited over the team's success, but he felt chagrined over not having had a direct hand in the creation of the working triode. He properly convinced Bardeen and Brattain to keep the discovery a secret from everyone, even Kelly, until they could successfully repeat the experiment—there was no use getting everyone excited if the result turned out to be a fluke. Replicating results is a key tenet of science. Within a week, though, Brattain was able to construct a full oscillator circuit that unambiguously proved the solid-state triode's signal-amplification functionality. Finally, thoroughly satisfied that their results were valid, they informed Kelly of their success.

Like Shockley, Kelly took the news with mixed emotions. He was miffed that his researchers had kept this secret from him for a whole week, but he was elated over their ultimate achievement. His vision had been proven correct and his twelve-year-long dream was finally a reality. Kelly could never expect a better Christmas gift. He, like Shockley, also ordered all his subordinates to secrecy for the time being as he conceptualized next steps.

THE ROLL-OUT . . .

To further develop the new technology, Kelly assigned more researchers and engineers to Shockley's team. Their goal: to initiate a solid-state triode prototype production effort, with the aim of reaching mass-production status as soon as possible. Other necessary supporting technologies, such as packaging technology for the device, were developed in parallel—given the finicky nature of semiconductors, the triodes would have to be shielded somehow; they could not be left to operate in open air, exposed to the elements.

Kelly notified the Defense Department of his team's discovery and work, hoping that this advancement would not be classified a military secret and thus be kept under lock and key. He was relieved when the military agreed that this technology should indeed be publicly investigated and applications sought throughout the world at large. After six months of intense work in secrecy, on June 30, 1948, Bell Labs held a special press conference in its New York City office to announce its new invention. The solid-state triode was officially named a "transistor," a name suggested by one of the Bell Lab scientists as a compression of *trans-resistor*, a partial description of the device's functionality. Using transistors, many new and compact prototype products were demonstrated at the press conference, including telephone repeaters, radios, and audio amplifiers. When reporters at the meeting listened to the music blaring from the speakers, they were surprised to find that they were not powered by the bulky, hot vacuum tubes so ubiquitous in amplifiers at that time. Instead, tiny metal cans housed specks of semiconductor material that, connected by thin wires, amplified the sound waves of music.

In spite of Bell Labs' announcement proclaiming that the transistor would usher in a new age, the world at large did not take much notice. In fact, the next day's edition of the *New York Times* hid its news item covering the unveiling all the way back on page 46! Clearly, this was a far cry from the sensation caused when Edison first demonstrated his lighting system in Menlo Park some sixty years prior. But perhaps that's no surprise. At the time of its introduction in 1948, nobody could imagine just how big an impact the transistor would make. Even experts of the industry could only predict that transistors might gradually replace vacuum tubes in telephone repeaters, radios and TVs, with some transistors perhaps required for the world's few computers. Nobody could have foreseen what the future really held.

Despite the lack of acclaim, Bell Labs saw that even the obvious market for end products utilizing the new transistor could be sizable. They avoided the common trap of underestimating the potential of new ideas and developments simply because of a lack of total understanding. Rather, they recognized that disruptive innovations often require time to germinate before they truly revolutionize the landscape and take over. Thus, Bell Labs' management soldiered on, directing Western Electric's subsidiary manufacturing operation to turn the transistor into a mass-producible product. At outset, production of the point-contact transistor (the common name for the Bardeen-Brattain invention) proved extremely difficult, not only because the device's basic design left many technical issues still unanswered, but also because the materials and fabrication technologies required for manufacturing remained nonexistent. Trying to control the many variables at every step of the manufacturing process was an almost insurmountable burden, and for many months, manufacturing yield was extremely low and performance of the transistors produced was poor. Luckily, at the helm of the project team stood an outstanding leader.

The joint Bell Labs/Western Electric team was headed by Jack Morton, an intense and capable technology manager. Under Morton's leadership, the engineers in the production department marched on relentlessly, solving countless technical issues and making improvements along the way. Their efforts reminded one of Edison's famous quotes: invention is one percent inspiration and ninety-nine percent perspiration. However, after a year of intense work, the team still could not reliably produce commercial-grade transistors, and people began to wonder whether or not this new invention would ever be manufacturable and financially viable.

. . . AND THE FIGHT

Right after Bardeen and Brattain successfully demonstrated the amplifying abilities of their transistor, Bell Labs immediately set about applying for a patent. Bell Telephone's early successes had derived heavily from owning the telephone patent, so the company's management always took the patent application process extremely seriously. The company's legal team took the lead on filing all the necessary paperwork with the U.S. Patent Office, and on the patent application document, they listed as the inventors of the transistor John Bardeen and Walter Brattain.

When Shockley saw this, he grew agitated and indignant. As the supervisor of the project and the man who had conceptualized its initial starting point, he felt that his name should be listed as an inventor, too. Brattain, in his blunt, cowboy style, made it clear that Shockley had played no direct role in the critical steps of the invention process and therefore did not deserve to be listed as one of the inventors. He couldn't reconcile the months of work he'd put in with Shockley's grandiosity, leading him to eventually, famously quip, "Oh hell, Shockley, there's enough glory in this for everybody!" Shockley stormed off, red-faced. This incident instantly sparked a deep rift between Shockley and Brattain. Bardeen, for his part, did side with Brattain, but he remained low-key about it, as usual.

Bell Lab's official release of the inventors of the transistor. Reprinted with permission of Alcatel-Lucent USA Inc.

In an attempt to placate Shockley, Bell Labs' management decided to apply for a separate patent based on Shockley's original field-effect concept introduced way back in 1939. In this way, they figured, he could reasonably feel as though he shared equally in the triumvirate's glory. However, during the due diligence process, patent attorneys discovered that a little-known schoolteacher in Canada had already applied for a similar patent some twenty years earlier, even though the idea had never been successfully demonstrated. This prior filing of a patent stymied any attempts to register Shockley's work as unique. When Shockley perceived that his own transistor research effort over the past decade was going to be for naught as his "subordinates" accrued all the fame and honors, he found it hard to swallow. Shockley had always been an arrogant person, and he was combative to an extreme. Thus, his conflict with the other inventors of the transistor only grew more and more heated over time.

Bell Labs' solution was to institute a requirement that henceforth, all photographs released to the media regarding the invention of transistor must show all three men together. The idea was that all three would appear to be equal partners in developing the device. However, this tactic failed in some respects, as the most widely reprinted photo depicting the invention of the transistor came to be one featuring Shockley seated at an experiment bench with Bardeen and Brattain reluctantly looking on from behind. They looked like a couple of students learning from their teacher, and given their previous arguments, this picture infuriated Walter Brattain, who found it very hard to live with for the rest of his life.

SHOCKLEY'S LAST LAUGH

Near the end of 1947, spurred on by the perceived grandeur being afforded to Bardeen and Brattain but not himself, Shockley resolved to sit down, marshal all his thoughts about p-n junctions, and come up with a solution even better than Bardeen and Brattain's point-contact transistor. He had for years held nebulous ideas about creatively using the p-n junction phenomenon to implement a solid-state amplifier, but during the war and the first years that followed, Shockley was tugged by too many distractions in his professional life to really pursue them. Despite the fact that he had never reached any systematic, specific conclusions, it appeared to Shockley that Bardeen and Brattain's point-contact transistor suffered from nearly as many technological and theoretical inadequacies as the point-contact rectifier had, despite its success, and there were enough pieces in place to design an improved device. Whether inspired by logic, drive, or jealousy, Shockley determined to do better.

Less than two weeks after Bardeen and Brattain showed him their point-contact transistor, Shockley shut himself up in a hotel room in Chicago and, from Christmas on through the New Year's holiday, wracked his brain dreaming of methods to employ p-n junctions rather than a point-contact design to make a transistor function. After much strenuous thought, a simple and elegant solution emerged in his mind. In his preliminary theoretical estimate, the concept appeared sound. There was still one looming technical question that required critical experimental data to validate, but Shockley felt he had a good start and resolved, for the time being, to wait.

After the New Year's holiday break, Shockley returned to Bell Labs. Kelly's dedicated team of solid-state researchers was hard at work refining the point-contact transistor just demonstrated by Bardeen and Brattain. The sprawling group held regular internal technical seminars at which scientists presented and shared their newest lab results. During one of these seminars, a researcher presented his experimental findings about the diffusion behavior of excess holes in n-type germanium. As fate would have it, this was the answer to Shockley's one

Shockley describing his invention: the junction transistor. Reprinted with permission of Alcatel-Lucent USA Inc.

remaining question about his own transistor design. At the end of the seminar, Shockley decided that it was time to speak up and reveal his concept of a p-n junction-based transistor. He abruptly stood up and, before everyone, provided a well-thought-out description of his invention and all necessary related theoretical analysis. His colleagues who were present at the seminar, including Bardeen and Brattain, simply stared at each other, flabbergasted. Shockley's striking ideas appeared both comprehensive and entirely sound, and they represented an impressive step forward in solid-state triode technology—yet he had never once discussed any of these thoughts with anyone! In two weeks, working alone, he had somehow managed to pull all of his scattered ideas together and create a masterpiece: a transistor design that did not rely on point-contact technology.

Shockley's revolutionary invention employed a three layer "sandwich" design with a very thin layer of p-type semiconductor bookended by layers of n-type semiconductor. (Note: in addition to this n-p-n transistor, he designed p-n-p transistors as well, with one layer of n-type semiconductor sandwiched by two layers of p-type.) If a tiny electric current, or signal, was injected into the middle p-type layer, an amplified signal current would flow from the top n-type semiconductor, through the narrow p-type layer, and into and out the bottom n-type semiconductor. The middle layer would function like a spigot on a water pipe: it could modulate the flow of water according to the amount it was turned—or in this case, the amount of current injected—or, at the extreme, it could turn the flow on or off entirely. Yes, the triode funneled electrical current rather than water, but the concept was similar. It functioned just like a vacuum triode. The only difference was that the entire functionality could be accomplished in a tiny piece of solid material!

The design Shockley proposed was simple and robust, making it very attractive to pursue. It didn't rely on using metal point contacts that were notoriously difficult to reproduce and control. Though the invention was purely theoretical at the time, Kelly nonetheless agreed to pursue Shockley's approach in parallel with the development of the point-contact transistor. Shockley also persuaded Kelly to allow him to file for a patent on this "junction transistor," or "bipolar transistor" as it was named, before even producing any experimental verification—a patent on which Shockley would, of course, be listed as the sole inventor.

By mid-1948, Shockley had completed the scientific theory of the p-n junction and his junction transistor. In addition to having it published in the prestigious technical journal *Physical Review*, he also leveraged it to produce an authoritative book on transistor physics entitled *Electrons and Holes in Semiconductors*, which became a classic treatise of semiconductor physics. Shockley's theories would prove paramount to the design of semiconductor devices and integrated circuits throughout the next sixty years. Speaking of this many years later, Shockley liked to point out one factor that spurred him on at the time. He called it "*the will to think*." We often have many ideas in our head but lack the determination to see them through. If, with enough will power, we mercilessly drive ourselves to bring our dreams to the fullest possible completion, we can accomplish many things in our lives. Of course, a fair bit of Shockley's will came from a dark and bitter source, but there is no denying his greatness.

Though Shockley largely worked alone, he did admire the work of Enrico Fermi, a physicist at University of Chicago best known for his demonstration of sustained chain reactions in nuclear fission. Shockley conceded that Fermi's thinking in statistics was a great inspiration to him, and when codifying his p-n junction theory, Shockley even granted an honor to the man by naming one of his technical terms an "imref," short for Imaginary Reference. Read backwards, imref has another meaning: Fermi! Though Shockley could be a bit of a curmudgeon, he was not above doling out credit or a joke on occasion.

THE ZEAL OF TEAL AND THE ÉLAN OF PFANN

While Shockley concentrated on building out transistor theory, preliminary experimental results proved the feasibility of his junction transistor concept.

However, mass production of the p-n junction-based transistor remained difficult because shortcomings in materials and fabrication technologies still presented major obstacles.

Gordon Teal. Courtesy of Texas Instruments, Inc.

One of Bell Labs' great competitive advantages was that it was loaded with talent. Top-notch experts in virtually every field were employed there, and their sheer number and technical diversity often accounted for the critical mass needed to tackle complex, multidisciplinary challenges. In Bell Labs' chemistry department, there was a particularly brilliant researcher by the name of Gordon Teal. His specialty was preparing high-purity, near structurally-perfect germanium and silicon crystals, and he supplied these high-quality samples to people like Shockley and Brattain for their experiments.

The standard process for preparing semiconductor samples called for first purifying the material in liquid form, known as "melt," and then cooling it down slowly to allow it to recrystallize, as in the case that led Ohl to discover the p-n junction. Germanium or silicon prepared this way would solidify into what is known as polycrystals. Such bulk material is comprised of multiple regions. Within each region, or "grain," the atoms are arranged in perfect order, known as single crystals. The polycrystalline sample contains multiple grains of varying atomic structural orientations, with boundaries between each grain, where structural defects and impurity atoms tend to accumulate. If a fabricated transistor were to contain these grain boundaries within its structure, its performance would be inferior, perhaps even nonfunctional. The ultimate solution to this problem was to develop material where the entire sample was single crystalline, with no grains and therefore no grain boundaries at all.

In 1949, Teal proposed to improve silicon and germanium crystal fabrication techniques by adopting and modifying a process invented during World War I by a Polish materials scientist named Jan Czochralski. This technology, known as "Czochralski pulling," had in the past been limited to use only in metals, but Teal felt strongly that the process might be applied to semiconductors as well. It might, Teal believed, even hold the key to producing large-scale silicon and germanium materials in total single crystal form.

Although Teal's proposal represented a potential major breakthrough, his request for funding was initially turned down by Shockley. Resolutely, Teal remained determined to go forward anyway, and he at last managed to find support from Jack Morton, the leader of the point-contact transistor manufacturing squad.

Teal did not squander this opportunity. Within a short period of time, he was able to successfully produce perfect germanium single-crystals without any grain boundaries. When these single-crystal materials were used to produce point-contact transistors, they showed significant and immediate improvement in both performance and manufacturing yield. Morton was ecstatic, and Shockley could only swallow his pride and admit that he had been wrong.

In the process Teal developed, a rod containing a small single-crystal seed material was first dipped slightly into a vat of molten germanium under precise temperature controls. Some of the melt would begin to precipitate to the end of the slightly cooler seed crystal. At that point, the rod would then be pulled slowly upward while rotating. If done correctly, the molten germanium would continuously solidify under the seed as the rod was pulled up, creating a single crystalline cylindrical ingot into which all of the molten material could eventually be incorporated. At the end of the process, the cylindrical ingot would be sliced into circular thin pieces, known as *wafers*, which were polished. These single-crystal wafers contained no grain boundaries, and were used to fabricate point-contact transistors.

Teal continued his research and soon realized that, should it be desired, he could intentionally add precise amounts of p-type and n-type impurities into the melt *during* the pulling process to produce uniform p-n junctions at circular cross-sections in the ingot. Through further refinement, he found that p-n-p or n-p-n sandwich structures could also be formed using this process, and that the thickness of the middle layer could be precisely controlled by the speed that the crystal ingot was pulled. Thus, his technique could not only be used to create Bardeen and Brattain's point-contact transistors, but Shockley's more advanced junction transistors as well.

In the winter of 1950, Teal produced the first junction transistor using this technique, and by the following spring, their performance, reliability, reproducibility, and yield had far outstripped that of point-contact transistors. More importantly, these devices behave exactly according to Shockley's theoretical predictions. Bell Labs could finally design and produce high performance transistors optimized for different applications, while these devices were also reliable and with consistent manufacturing yield.

In addition to Teal's breakthrough in producing structurally perfect single-crystal semiconductors, Bell Labs concurrently developed another important materials technology breakthrough known as "floating zone refining," which allowed the purity of the single-crystal semiconductors to reach unprecedented levels. This process hinged on the fact that as semiconductors melted, most of their impurities tended to physically remain within the melt region rather than return to the solid host. In the floating zone refining process, a solid rod of germanium was first mounted at both ends in a specially designed furnace with a movable, concentric coil heater that completely engulfed a narrow section of the rod like a sliding belt. The coil was inductively heated to a temperature just above the melting point of germanium, whereupon it melted a thin cross-section of the rod while the rest of the material remained solid. The melted zone would be held in place, or "floated," by surface tension on both interfaces. Impurity atoms would stay in this melt zone, and as the coil slowly moved across the rod, more and more impurity atoms would be "swept" into the moving melt zone. As the coil moved slowly back and forth from end to end, most of the impurities would end up segregated into either end of the rod, leaving the sample in the middle nearly free of impurity atoms. Via the floating zone refining process, impurities in germanium could be reduced to below one part in ten billion—in other words, the germanium would emerge from the zone refining process an astonishing 99.9999999% pure.

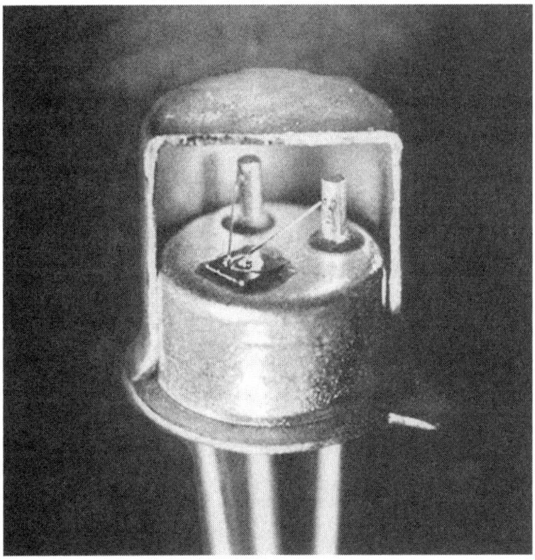

Cross-section of packed transistor chip in a small metal can. Courtesy of Fairchild Semiconductors

The inventor of the ingenious floating zone refining process was a man named William Gardner Pfann. Pfann began working in Bell Labs' mail delivery room straight out of high school.

He was good with his hands and a fast learner, and when he began taking night classes in chemical engineering at nearby Cooper Union, he transferred from a mail delivery clerk's job to that of a junior lab assistant. Pfann was instinctive, reliable, and capable, and soon, many senior researchers sought out the young man's service. Gradually, Pfann developed a reputation as a problem solver, and by the time point-contact transistors were invented, Pfann had become a part of the extended team Kelly mobilized to rapidly productize the new technology.

Pfann's original assignment was to develop a reliable approach to packaging the tiny transistor chip. His solution was a hermetically sealed, tiny metal housing, into which ran fine gold wires soldered to the transistor's contact pads. This invention was adopted, and the design remained widely used in transistor packaging from the 1950s through the 1970s. Still, the impact he made in developing the floating zone refining process was even greater, a true masterpiece in materials technology.

RESOLUTION

In June of 1951, Bell Labs held another press conference in New York City to launch its new junction transistor. This meeting occurred three years to the day after the first announcement of the point-contact transistor. At the subsequent reception, Shockley was the focus of attention—for their part, Bardeen and Brattain did not even attend.

Not long afterwards, Bardeen left Bell Labs and joined the faculty of University of Illinois to engage in theoretical research on superconductivity. He twice received the Nobel Prize in Physics: once in 1956 for the invention of the transistor, and again in 1972 for his contributions to superconductivity theory. Historically, there haven't been many scientists as brilliant and productive as Bardeen, and yet he was an extremely humble man. He died quietly in 1991 at the age of eighty-three.

For his part, Brattain stayed at Bell Labs until his retirement in 1967. He later moved back to his hometown in the Pacific Northwest to teach. In his later years, Brattain found it very irritating when youth of the 1970s and 1980s used boom boxes—powered, as fate would have it, by transistors—to play deafening rock 'n' roll music in public. He would say tongue-in-cheek that he regretted ever having invented the device. Brattain died in 1987 at the age of eighty-five.

Shockley, for his part, remained in the electronics industry even after inventing the junction transistor and enjoying the subsequent success of completely replacing Bardeen and Brattain's invention. And yet Shockley, ever a man of pride and action, came to find himself in short order in the center of yet another all-consuming corporate controversy.

14

LAUNCHING THE ELECTRONICS INDUSTRY

SHARING TECHNOLOGY

Thanks to the work of Shockley, Teal, Pfann, and many others, the junction transistor entered mass production. Though the core technical issues had been solved, many business questions remained. What would be Bell Labs' next step with the transistor? How would they go about commercializing transistor technology? Bell Labs' business model was different from that of private sector, for-profit enterprises. It was a part of a government-sanctioned, regulated monopoly. The working capital used for its research came via subscriber billing surcharges that were approved and authorized by the government. Though the transistor had clear and useful applications in the company's core telephone business—for example, by improving both the quality and operating cost of long distance telephony and network switching—the commercial value and potential impact of the technology was much beyond AT&T's business domain. However, Bell Labs' parent company, AT&T, was not in the business of selling electronic components. Another important but unspoken factor was that even if AT&T did generate windfall profits from capitalizing the transistor technology in the marketplace, such profits would only induce government regulators to force the company to lower its telephone rates in order to maintain its preapproved profit margin.

Lastly, there was also some whispering that the federal government might force AT&T to spin off Western Electric—or even remove the umbrella of protection over its monopoly—in order to increase competition in the telephone business. Thus, there was no benefit for AT&T to try to hold on to the transistor technology for itself or to lord it over the marketplace as other companies might—indeed, there were a number of negatives to such an approach.

In the end, the management of Bell Labs, with concurrence from AT&T's top management, decided to share the technology with the global marketplace and broaden the transistor's impact. In exchange for a modest fee of $25,000 for manufacturing, production, and licensing rights, companies from any nation in the NATO Alliance—the late 1940s and early 1950s marked the nascent stages of the Cold War, after all, and AT&T was highly attuned to political realities—could purchase the right to utilize transistor technology. For many farsighted enterprises, this would prove to be a once-in-a-lifetime opportunity. In a totally unplanned way, Bell Labs, with its vast technical capabilities and its unique business arrangement, turned out to be the perfect model for an R&D laboratory: developing the right cutting-edge technologies and then disseminating the knowledge unselfishly to benefit the rest of the industry.

In April 1952, more than one hundred engineers from forty-odd companies around the world gathered at Bell Labs to receive ten days of training in transistor technology. Bell Labs was sincere in its efforts to transfer all of their knowledge, and they held detailed lectures and breakout sessions explaining the core physics of semiconductor devices, the principles of transistor design, drawings and descriptions of Teal's crystal-pulling equipment, and information about Pfann's zone refining process, along with other valuable information. The forty-some participating companies in attendance included big corporations like GE, IBM, RCA, Siemens, and Philips, as well as still mid-sized companies like Motorola, Sylvania, Philco, and Raytheon. There were even some small, then little-known companies like Texas Instruments. Just as the Moore Lectures held in the summer of 1946 helped spearhead the beginning of the computer industry, this series served as the launching event for the global semiconductor industry.

NEW PLAYERS

Bell Labs carried out these technology transfer sessions three more times in 1952. One participant at a later event was a small Japanese company called Tokyo Tsushin Kogyo (TTK). Though scarcely anyone outside the company had ever heard of TTK before and no one could have predicted it, TTK would later become one of the most profitable and impactful companies in the world, and the foundation of their success was largely due to their licensing of the transistor technology from Bell Labs. In fact, it is worth noting that throughout the spectacular growth of the transistor-fueled electronics industry, none of the major corporate entities did particularly well in the semiconductor business. Most of them were deeply ingrained with their existing, highly profitable vacuum tube businesses, and the new transistor was always viewed as an unwelcome competitor by existing technical and management staff. The most successful companies were the ones who were not encumbered by past history, such as Motorola, Texas Instruments, and Japan's TTK. Such is often the case: as with Western Union's early dismissal of the telephone, organizations that enjoy great success with an incumbent technology can become institutionally blind to new opportunity and lack the commitment needed to succeed in an evolving marketplace with disruptive new technologies.

Motorola began in 1928 as a manufacturer of car radios. Even its adopted company name was derived from this product: "motor" referred to the car, and "-ola" was a common brand name for audio products at the time, for example RCA's Radiola radio. Later, it developed a two-way wireless telephone for use in police cars. Building on such products, Motorola successfully produced many types of portable military radio communications equipment during World War II, including the ever-popular walkie-talkie.

After the war, Motorola's radio and radio-derivative businesses grew rapidly. Transistor technology was highly complementary to the company's core portable communications business, so at Motorola, the semiconductor division grew rapidly and with the benefit of great corporate resources. What's more, the company was just small enough that it could give its transistor team the fostering and care it needed to shine.

Texas Instruments, or TI, was founded as Geophysical Services Incorporated. Its original products consisted of several models of sonar instruments used in exploring oil fields, a key industry in Texas. In the early postwar period, its annual gross revenue was a paltry $3 million. But in 1946, TI gained a new leader in Pat Haggerty. Haggerty was trained as an engineer. He was also a visionary who pioneered the art of "strategic management." Haggerty was constantly on the lookout for the company's breakout opportunity, and he quickly became convinced that Bell Labs' new transistor technology represented just that. After licensing the technology from Bell Labs, TI committed all of its limited resources to building up a unique business culture focused on incubating the transistor business. Within the next thirty years, TI grew to become the world's single largest semiconductor company. Even to this day, TI remains a key player in the global semiconductor business.

For its part, TTK achieved the greatest success of all of these companies, though it never became well known under its original name. TTK's story would unfold over the course of time.

THE DEBUT OF SILICON

Starting in 1953, many companies began to produce transistors based on the Bell Labs technology, and thus, a new industry was born. Of course, initial production rates were low and fixed costs were high, so as with any other new technology, transistors were very expensive. A typical transistor would sell for around $20 in the mid-1950s, while a vacuum triode of comparable performance would cost less than $2. But the unique advantages of transistors—low power consumption, high reliability, long lifetime, and compact size—were critical for many applications in the military, where price is nearly always of secondary importance to functionality and performance. As the Korean Conflict ended and the Cold War heated up, the demand for ever-better weapons systems performance brought along with it demands for cutting-edge solid-state electronic components. Transistors began to take the place of vacuum triodes in most of the U.S. military radios, radar systems, and computers, and it was the patronage of the defense industry that sustained the critical early growth of the transistor industry.

Even though Teal demonstrated the first junction transistor in 1950 using the crystal-pulling technology, this approach was expensive and not amenable to low cost mass production. To reduce cost, researchers at GE developed a new "alloying transistor" technology based on germanium, which was licensed to RCA and other companies, and soon it reached mass production status. However, the germanium alloying transistor was only a transitional product, and starting from the mid-1950s, they were replaced by a far superior technology based on silicon.

In 1954, two new and important technological innovations were developed. The first was gas phase impurity diffusion technology. By intentionally diffusing gaseous "impurity atoms" such as boron or phosphorus into germanium at a high temperature, these p-type and n-type impurity atoms could be precisely incorporated into germanium single-crystal wafers with controlled concentration and penetration depth, allowing Bell Labs to fabricate uniform, large area, high-quality p-n junctions across wafer surfaces. Eventually, researchers pushed toward a still more refined, multistep diffusion process that could successfully produce full n-p-n and p-n-p junction transistors in this same way. These advances drove down transistor production costs by more than an order of magnitude over Teal's crystal-pulling technique, and they were also superior to GE's alloy transistor technology in terms of performance uniformity and production yield.

In addition to lowering costs, the impurity diffusion process was ideal for creating large, uniform p-n junctions across an entire wafer, and these large area p-n junctions proved to be suitable for another new and important application. Recall that in 1941, Russell Ohl discovered that his inadvertently created

Solar photovoltaic panel used to power remote telephone repeaters (1956). Reprinted with permission of Alcatel-Lucent USA Inc.

p-n junction was sensitive to light. When the semiconductor absorbed light, it generated an electrical current flowing across the p-n junction, converting light energy into electrical energy. This phenomenon was known as the photovoltaic effect. Gerald Pearson, a senior experimental physicist in Shockley's group, quickly realized that by exposing these large area p-n junctions to sunlight, he could convert solar energy directly into electrical power.

Because of its smaller energy band gap, germanium was not well suited for converting solar energy to electricity; however, silicon was near perfect for the task. In 1956, Pearson and his team adopted and modified the impurity diffusion process from germanium to silicon and demonstrated the first large-area, single-crystal silicon solar photovoltaic cells. The conversion efficiency of these first silicon solar cells reached over 6 percent, which provided adequate power to remotely located telephone repeaters mounted atop telephone poles. It was a clever application, and with the introduction of this product, AT&T launched the age of photovoltaic solar energy.

Around the same time Pearson unveiled the solar cell, yet another major breakthrough in transistor technology emerged—but for once, it did not take place at Bell Labs. In 1952, TI licensed Bell's transistor technology and began mass producing germanium transistors. The company's strategic business plan, as elucidated by its visionary leader, Pat Haggerty, was to first master the basic transistor technology, then to develop TI's own capabilities that were both unique and superior to that of its competitors. Under Haggerty's leadership, TI set up a special laboratory to drive this strategy forward.

The success or failure of scientific innovation nearly always hinges on the quality of research staff. Haggerty knew this, so he set about identifying and recruiting the best talent as a top priority. Right at the top of his list was Gordon Teal of Bell Labs, the man who revolutionized semiconductor materials technology and the junction transistor production process. Haggerty saw in Teal an exceptionally creative researcher, a strong leader, and a man who possessed complete firsthand know-how about every detail of transistor manufacturing technology. Haggerty also foresaw that since Bell Labs was licensing out its core semiconductor technology, the company would likely not be expending significant additional resources to further refine the technology, which might make Teal amenable to changing employers. So Haggerty reached out to recruit Teal. Lucky for Haggerty, Teal, himself a Texan, was indeed excited by the opportunity to remain at the forefront of transistor research and lead his own product team, and in 1952, Teal left Bell Labs to return to his home state.

After Teal joined TI, he quickly resumed trying to improve existing germanium transistor production methods. More importantly, he also launched a project to develop a transistor manufacturing technology based on silicon, which he had already begun to research while still employed at Bell Labs.

Up to that time, all transistors had been made from germanium because it was far easier to work with. Germanium's melting point is comparatively low, just above 900° C, while silicon's is over 1,400° C. Many manufacturing technologies, such as floating zone refining, could be used for germanium but not for silicon—the zone refining furnaces just couldn't get hot enough to melt the silicon material. However, the idea of silicon transistors was extremely appealing, despite the difficult problems their fabrication posed. To begin with, since silicon has a larger energy band gap, silicon transistors can operate at higher temperatures than germanium. Typically, germanium transistors could perform only at temperatures up to about 70° C, but silicon transistors could function beyond 120° C. Since transistors heat up during operation, these enhanced operating temperature limits implied that silicon transistors would find many more applications and markets than germanium ones could. Additionally, while germanium sourcing was expensive, silicon was and is cheap—silicon can be extracted from sand; there is an inexhaustible supply of the element on earth. Additionally, silicon is nontoxic, chemically stable, and mechanically strong and has excellent thermal conduction properties. All these properties made silicon the material of choice for transistors, at least in theory. (There was one more extremely important advantage in silicon's favor that was not realized at the time but turned out to be crucial to the success of integrated circuits. This subject will be discussed in later chapters). The opportunity to create the world's first silicon junction transistor was a huge draw to Teal as he changed jobs, and in 1954, Teal delivered astounding results.

After working quietly and under strict confidentiality at TI for over a year, Teal was finally able to successfully fabricate the world's first silicon transistor. He announced his creation to the world in a most unusual way. While attending a technical conference, Teal watched as one speaker after another extolled the virtues of a theoretical silicon transistor and then lamented the frustrations of being unable to solve the requisite fabrication challenges. When it was at last Teal's turn to speak, he simply walked to the podium, played music from an audio amplifier made with a germanium transistor, then dropped the transistor circuit board into a glass of boiling water. To no one's surprise, the music went dead, because the germanium transistor could not function at 100° C, the temperature at which water boils. Teal then casually took out from his pocket a replacement transistor, pressed play again, and then immersed *that* circuit board in the boiling water. This time, the music kept playing. Slowly, it dawned on the audience what exactly they had just witnessed. As the assembled crowd broke into uproarious applause, Teal dramatically announced TI's spectacular success in ushering in the new age of silicon electronics.

THE TRANSISTOR RADIO

In the early years, demand for transistors came mainly from military markets. Even though unit prices were high, overall market volume was relatively small, so manufacturers did not have an easy time reaching profitability. The key to expanding the business was to replace vacuum tubes with transistors in products in other large markets, and at the time there were two: computers and consumer electronics. For the computer market, steady progress was already on track; as for the consumer market, the outlook was still unclear.

TI's Pat Haggerty, ever ahead of the curve, was one of the first industrial leaders to recognize the importance of opening up consumer products to transistors, and he envisioned the radio as the most natural point of entry. His plan was to use transistors to develop miniaturized radios powered by batteries. Unlike Armstrong's first custom-made "portable radio" that he brought to picnic on the beach in 1922, these transistorized radios would be small and lightweight enough to be carried around in pockets. This was a revolutionary idea with great market potential. TI wouldn't need to convince consumers to invest in a whole new form of entertainment—radio had already long been popular. All the company had to do was convince buyers to try to enjoy a favorite pastime in a new, more accessible way. What's more, the existing radio market was big enough that by capturing even a tiny slice, TI could ensure its success.

Haggerty's business instincts were impeccable, but he had a problem: TI's business niche was in transistor components; prior to that, it had serviced only the oil industry. The company had no toehold in the consumer market. TI needed a business partner.

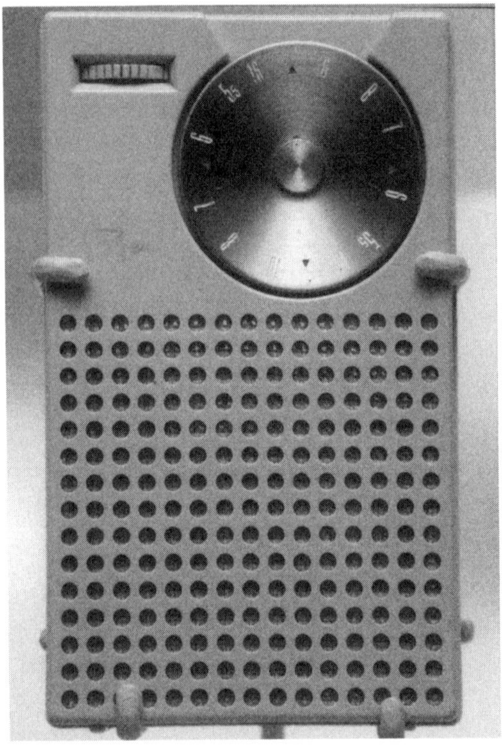

TI Regency TR-1 portable transistor radio. Courtesy of Texas Instruments, Inc.

TI approached many well-established radio companies, but its proposal was not taken seriously. RCA and several other major radio companies had doubts about the price and performance of transistors—they either didn't believe Haggerty's vision that the market could support such a product, or they couldn't muster the institutional buy-in needed to overcome inertia and attempt something so novel. It took a fair bit of searching, but finally, TI found a partner in IDEA, a medium-sized, technically innovative radio company. While the other, larger companies were too risk-averse and too concerned with simply hanging on to their slices of the pie to make a gamble in the marketplace, IDEA was willing to dive right in and take a shot—a potentially very lucrative shot—at revolutionizing the industry.

Working together, TI and IDEA produced the first pocket-sized, consumer-oriented transistor radio, which they branded and sold as the Regency TR-1. The radio was manufactured with IDEA's patented new circuit board process, using four TI-produced germanium transistors. TR-1 was priced at $49.95, or about $600 by 2012 prices. Though it was by no means an outright money loser—manufacturing technology had improved in the intervening years, and transistors could now be fabricated for closer to $8 apiece rather than $20—this price point still provided a net negative sales opportunity to TI. In 1954, TI could sell germanium transistors to the military for $16 each, so the four transistors used in a radio could be sold individually for $64—higher than the price for which TI and IDEA sold their entire radio! However, the two companies agreed that they could not price the radio beyond $49.95; customer feedback implied that this was probably the highest price that the market would support. So the only option they had was to significantly reduce the cost of the product, especially the transistor.

The Regency TR-1 was heavily promoted during the 1954 Christmas shopping season. Although it was expensive, Haggerty was right: people wanted it, and demand far outstripped supply. No less a buyer than the president of IBM bought 300 of the radios, presenting them as gifts to his company's most outstanding employees. He challenged his staff to learn from this small, innovative company in Texas and to wean away from using vacuum tubes in future IBM products. Coincidentally, TI always enjoyed a close working relationship with IBM, and IBM was TI's biggest customer for many years after that.

Despite its success in the marketplace, the TR-1 transistor radio itself did not generate profit for the company. Even as the unit cost of manufacturing the germanium transistors decreased dramatically in subsequent years, TI still had trouble making profit on the TR-1, as more low cost competitors entered the market. Even though the company enjoyed the uptick in sales revenue, and even though Haggerty had been proven conceptually correct, TI's management grew increasingly suspicious that this might not be the right strategic direction for the company to pursue. Traditionally, TI's core business lay in high-end military and industrial electronic instrument products. The culture and business philosophy required to play in the consumer electronics space were totally different. In order to succeed in both markets, the company would need to operate with two competing and opposing business cultures simultaneously. For any company, this represented an extremely difficult challenge to handle, one that could never be easily tackled even with the best management talent.

Eventually, TI was forced to downgrade its emphasis on the radio and its joint-venture partnership with IDEA. In later days, TI would try to reenter the consumer electronics end-product market, but generally speaking, its efforts were only marginally successful, mainly because it was difficult for TI to maintain a long term, sustainable competitive advantage over a horde of competitors. Many big high-tech and aerospace defense companies have tried to leverage their technology strength to diversify into large commercial markets over time; however, this strategy has rarely worked, and examples of success are few and far between.

After seeing the high sales volume and ready market adoption of the Regency TR-1, other companies followed in TI's footsteps and introduced transistorized niche products with higher margins than the radio, including hearing aids and desktop calculators. Compared to products made with vacuum tubes, these new transistorized products were much smaller, lighter, and more reliable. Consumers loved the new technology, and the general public could finally directly acquaint and experience themselves with transistors. The only shortcoming of these new products remained their expensive price tag.

JAPANESE PIONEERS

Among the attendees at Bell Labs' 1952 transistor technology information sessions were representatives from a small Japanese company called Tokyo Tsushin Kogyo,

Ibuka (left) and Morita. Paris Match via Getty Images

or TTK. The company had been founded by two young entrepreneurs in Japan named Akio Morita and Masaru Ibuka, who had gotten to know each other while serving in the Japanese Navy near the end of the Second World War. Ibuka was an outstanding product design engineer and a solid manager, while Morita was something of a renaissance man, versed in many different, important disciplines.

Morita had been groomed as heir-successor to a family-owned sake brewery of some renown since the age of thirteen. The brewery, based in Nagoya in central Japan, had been in the family for fourteen generations, and Morita's father—much like Marconi's—had expected the boy to follow in the footsteps that lay before him. Indeed, Morita had shown instinctive and creative business acumen from a young age. But Morita also had a broad range of personal interests, including a fascination with radios. His formal university training was in electrical engineering, and after the war, though his family's business had not been damaged, Morita decided to strike out on his own to launch a small radio and tape recorder company in Tokyo with his friend Ibuka. The men named their venture TTK: Tokyo Tsushin Kogyo, or in English, the Tokyo Telecommunications Technology Company.

Within three years, the company successfully introduced a compact tape recorder to the consumer market—naturally, the recorder utilized vacuum tubes, as was then the standard. Quickly, TTK built up a niche for its products. But the two ambitious young men were not satisfied with their success, and they remained on the lookout for ways to propel their business farther forward.

Morita paid close attention to the development of the transistor from the time Bell Labs first announced its invention in 1948. In addition to pure technological curiosity, he also vaguely recognized that this new technology might have an impact on TTK's future. When in 1952, Bell Labs decided to broadly license the transistor technology, Morita made a resolute and daring strategic decision: TTK would officially license it, becoming the only Asian company to do so. They raised the $25,000 fee and applied for permission, and they were elated to find that they received U.S. governmental approval.

Morita and his small technical team travelled to America for the first time in 1952. It was an eye-opener. The new world was filled with fresh ideas and opportunities, and Morita became convinced that the time to act was upon them. His engineers attended the Bell Labs sessions, and upon returning to Japan, Morita and the TTK team quickly reproduced Bell Labs' know-how and commenced production of their own germanium transistors. The product was good, but to their dismay, they didn't make much money. The problem was that demand for transistors in Japan was very low, principally because the domestic economy of the early postwar era did not possess developed, modern computer or military electronics industries, the two sectors that had supported the transistor industry in the United States during its startup years.

With a license and know-how but no market in which to play, TTK spent the next two years simply surviving. But 1954 saw the release of TI's Regency TR-1 transistor radio, and when Morita had the opportunity to study the product in detail, he immediately realized that perhaps this was the great business opportunity he'd long been waiting for. Not only could TTK already produce transistors, but compact radios and tape recorders were the company's core business. What's more, TTK was not likely to encounter the business dilemma of trying to play in two business spheres at once, as TI had. Morita and Ibuka knew the ins and outs of this business. In the transistor radio, they believed they had a real shot at growth and profitability.

In 1955, Morita came to America again; his goal this time was to assess the transistor radio market. Just a bit of research convinced him that the market was very big indeed. Estimating based on Japanese manufacturing costs—which were significantly lower than those in the United States and also included a favorable rate of exchange for exports—Morita came to believe that selling a TTK-branded radio could reap an attractive profit. However, TTK did face a roadblock that TI did not: capital. TTK was not nearly as well-funded as TI had been; to produce a new transistor radio, TTK would have to single-mindedly commit all its resources and bet the company. But Morita and Ibuka believed in their product and their idea, and they were willing to take the risk. Provided all their energies and resources remained single-mindedly focused to ensuring product performance, quality, and cost control, they believed they could emerge from this labyrinth victorious.

Based on the company's business plan, TTK was able to raise the capital in debt markets and begin development of its transistor radio. But before launching the new product, Morita had the foresight to recognize the importance of *branding*. He strongly felt that to successfully promote the company and its products around the world, he would need a name with cachet. Tokyo Tsushin Kogyo, much as he liked it, would not cut it. The "Tokyo," especially, held negative connotations; in those days, the phrase "Made in Japan" served as a signifier for cheap and low-end goods. After several discussions, Morita and Ibuka agreed that the company needed a shorter name, one that would be easily remembered in English. As the product was an audio device, they began with "sonus"—Latin for "sound"—and blended it with "sonny," an American slang word that seemed to fit with the times. Thus, their new brand name was born: Sony.

In 1955, Sony released its miniature transistor radio, model TR-55. IDEA attempted to sue Sony for patent infringement, but they underestimated the willfulness of this small Japanese company, and the case was not followed through. Indeed, the TR-55 was very similar to the TR-1 in both appearance and in performance. The main difference between the Sony and the TI/IDEA models was price: the Sony radio sold for just $29. Not only was this price significantly lower and therefore more attractive to the marketplace, it was also a price point that IDEA and TI could not match, but at which the low-overhead Sony could still enjoy a healthy profit margin. With further improvements, Sony introduced model TR-63 in 1957, and on the strength of this much-improved transistor radio, Sony conquered the world.

Sony's portable radio success story marked the first in a long line of electronic products that were first developed by American companies and later perfected and cornered by Asian companies. This pattern has subsequently repeated itself many times, the underlying reasons for which we shall discuss later on in the book.

THE TRANSISTOR ERA BEGINS

In 1956, AT&T successfully completed its first transatlantic telephone cable. Despite the telephone being invented just thirty years after the telegraph, the first intercontinental, undersea telephone line wasn't completed until almost one hundred years later than its telegraphic predecessor, mainly due to technical reasons. The telegraph is based on digital signals that can be forwarded by simple relays; telephones, on the other hand, use analog signals and require multiple repeaters with high-performance amplifiers in order to preserve their sound fidelity over long distances. As long as vacuum-tube triode amplifiers were used in telephone repeaters, their high power consumption and low reliability rendered them largely unsuitable for the undersea environment. But amazingly, the first generation transatlantic telephone cable actually used vacuum tubes! It wasn't until 1959 that these tubes were replaced by transistors and true, reliable transatlantic telephony became a reality.

1956 was also the year that Bardeen, Brattain, and Shockley jointly received the Nobel Prize in Physics for their contribution to the development of the transistor. By inventing the point-contact transistor, Bardeen and Brattain proved to the world the feasibility of a solid-state amplifier. Shockley, for his part, followed that creation with the invention of the practical junction transistor and the development of the fundamental semiconductor device physics that would guide the growth of the entire electronics industry.

Of course, the development of the transistor involved the contributions of a large group of brilliant people, not least of who was Mervin Kelly, the man behind the scenes upon whose farsightedness and support all of this work stood. The mere existence of Bell Labs, too, played a crucial role in this success—during this period, all of Europe and Asia were still rebuilding from the devastation of the Second World War, and nowhere else in the world was there such a concentration of outstanding technical specialists, high-quality facilities, and innovative spirit. Bell Labs' resources, unique operating model, focused research objectives, and incomparable technological capabilities paved the way for some of the greatest advances of the era, both in hardware and in information sciences. Among them all, the invention of the transistor was the single most pivotal. This marks an important step forward for mankind in its continuous effort to harness the power of the electron. Some historians would even compare the importance of the invention of the transistor to that of the wheel. Just as the wheel revolutionized transportation, transistor technology has fundamentally enabled the information age that we all live in.

15

THE DAWN OF SILICON VALLEY

WALL STREET JOURNAL OR *PHYSICAL REVIEW*?

By the late 1950s, the transistor had assumed its rightful place at the forefront of the technological revolution. It had already found homes in consumer electronics and communications systems, and a critical role in powering the new computer industry was clearly on the way. After all, if the super-powerful ENIAC had been limited principally by the temperament and short operating life of vacuum triodes, it was easy to imagine that a similar machine employing transistors could be the most potent and reliable computer ever created.

Though the prospects of a new electronics industry driven by transistors appeared bright and limitless, the sad fact was that the most productive and creative time for semiconductor research at Bell Labs was over. Bell Labs had developed the transistor and licensed out the core technology; as Haggerty had intuited when courting Teal, the organization had no need to continue to sink huge funds into performing basic semiconductor research. The door to the coop was left wide open; it was only a matter of time before most of the birds would leave the nest.

The first to leave was Teal in 1952, followed by Bardeen in 1954. Shockley, ever out for adventure—ever willing to scale a cliff, or speed in his car—was tempted to follow this exodus.

In 1954, Shockley turned forty-four, and he began to go through a mid-life crisis. He traded in his beloved MG sports convertible for an elegant Jaguar sedan. Shockley was already famous in the world of scientific research; however, he wanted not only fame, but also fortune. The problem was, there was no path to getting rich at Bell Labs. Bell Labs owned all the technology its employees created; all inventors got was a symbolic cash award of $1.00 for each patent they filed. Career-wise, the prospect for Shockley to move up the ladder at Bell Labs seemed dim: he knew that he was technically respected by his colleagues, but he was not well liked as a manager or a coworker. After some deep soul searching, he decided it was time to say goodbye to Bell Labs and the researcher's life. The peak of basic semiconductor research had, he believed, already passed. The next logical step would be the rapid rise of the new semiconductor industry, with pioneers exploring the uncharted unknown.

Late in 1954, Shockley left Bell Labs and signed on as a Visiting Professor at Cal Tech, his alma mater. He didn't last long there, though—after all, academia is no place to seek one's fortune. From there, Shockley progressed to serving as a senior consultant for the Department of Defense, but that still didn't quite fit the bill. Finally, Shockley decided to launch into the world of business. He no longer felt motivated to solve purely technical problems or publish scientific papers; it was money and business success that he was after. As he intimated to his associates, he'd had enough of seeing his name in prestigious but largely academic publications like the *Physical Review*. He wanted his name in the *Wall Street Journal*.

As Shockley began to evaluate his options, it became clear that the best chance for him to achieve his new goal in life was to start his own company to produce cutting-edge semiconductor devices. Although he knew almost nothing about business, he did know for sure that he wanted to be his own boss; thus, joining any of the existing semiconductor companies was out of the question. He did realize that to start a business, one must first attract investors, and in this vein, Shockley had a leg up: he possessed a great reputation and was regarded as the top techni-

Arnold Beckman. Courtesy of the Archives, California Institute of Technology

cal authority in this nascent field. Upon hearing of his intentions, several powerful potential investors, including John D. Rockefeller, Jr., approached Shockley with the aim of supporting his venture. However, they all backed away after discussions with Shockley revealed his dearth of business skill and his inflexibility. Shockley was near throwing up his hands in exasperation when he thought of a friend that he'd met at Cal Tech while he was an undergraduate: a successful businessman by the name of Arnold Beckman.

Arnold Beckman was an outstanding scientist with a PhD in chemistry and a strong background in electronics. He actually even worked for several years at Western Electric's Technology Department, the precursor to Bell Labs. He had been a graduate student at Cal Tech by the time Shockley came along as an undergraduate. He was also, as it turned out, a very successful entrepreneur.

One of Beckman's friends owned a Sunkist lemon cooperative in Southern California. Once it came time to squeeze and preserve the lemon juice, this friend needed a method to continuously measure the acidity of the concentrate and additives. However, there was no commercially available instrument to do that at the time. In response to that need, Beckman invented a successful—and ingenious—electronic pH meter. From there, he went on to build Beckman Instruments, a company that specialized in instruments for chemical, biomedical, and optical research. By 1955, annual revenue for the company had reached $20 million.

Beckman was a sincere and unassuming—and wealthy—gentleman with a particular zest for science and technology. Shockley approached him for support, and, after long and hard consideration, Beckman decided to support his friend with an investment of $1 million to establish a commercial silicon transistor product company called Shockley Semiconductor Laboratory. Beckman hoped to set up the company close to Cal Tech and the headquarters of his company in the Los Angeles suburbs. But Shockley was fixed on establishing his business in Palo Alto, just south of San Francisco in Northern California. The climate was pleasant, and with Stanford University and UC-Berkeley nearby, Shockley argued, being in Palo Alto would allow him to best remain in touch with the field's newest scientific and technological advances. Frederick Terman, Stanford University's visionary dean of the School of Engineering, was also an indispensable source of support, and he lobbied hard on Beckman as well. Finally, Beckman relented.

Beckman might not have realized that regardless of the reasons he gave, much of Shockley's desire to establish his business in Palo Alto came from the fact that Palo Alto was Shockley's home town; his aging mother still lived there, and he wanted to return to his roots. But what nobody realized—not even Shockley—was that this decision to base the company in Northern California rather than Los Angeles would lead directly to the birth of Silicon Valley.

SHOCKLEY AND THE TRAITOROUS EIGHT

In 1955, the Shockley Semiconductor Laboratory was officially established, and Shockley mounted an intense effort to recruit the best talent in the country to work for him. It was a bitter disappointment to him that none of his old colleagues at Bell Labs would agree to join him, but then, it should have come as no surprise. Everyone who'd ever worked with him knew that Shockley was an extremely difficult person to get along with, and, given the choice, most preferred to keep a respectful distance. So Shockley turned his attention to recruiting fresh, young technical talent from other companies all over the country, as well as new PhD graduates from top-tier universities. This, too, turned out to be a decision whose impact would be felt throughout future generations—even today, a youthful spirit pervades both the Silicon Valley region and the high-tech industry.

Shockley's reputation was at its zenith in the mid-1950s, and many young scientists considered it a personal honor to be asked to join Shockley's laboratory. Shockley, for his part, had good instincts when it came to judging and recognizing technical talent, and he soon built a strong and energetic team consisting mainly of men in their late twenties and early thirties, stocked with specialized technological expertise. Among these men was Robert Noyce, a promising, young research engineer from Sylvania with a PhD from MIT. Another was Gordon Moore, a highly intelligent chemist from Cal Tech. The members of this young team willfully relocated from all over the country to Palo Alto, excited to have the opportunity to work with the world's top authority in transistor technology and together build a company that would dominate the new industry. That enthusiasm was further bolstered when, the year after the company's founding, Shockley received the Nobel Prize in Physics. The whole team was engulfed in eagerness and glory.

Unfortunately, the happy days at Shockley Semiconductor Laboratory did not last. In conducting scientific research, Shockley was brilliant and peerless, but when it came to managing businesses and people, he was easily among the worst men for the job. He was irrational, dictatorial, and insensitive in the extreme; he also suffered from a serious case of paranoia. All these shortfalls, however, might have perhaps been overcome had Shockley not also lacked even the slightest inkling of market needs, product strategy, or team focus. Not long after the company commenced operations, he shifted the focus of product development from the transistor to the four-layer diode, another technology that he'd invented at Bell Labs. This diode—basically a controlled electronic switch—consisted of four separate layers of p- and n- semiconductor material. Its specific capability was that after switching a current signal on or off, the four-layer diode could stay in its position without consuming much additional power. Theoretically, this made it ideal for telephone central switch applications, which was the original objective that Kelly had wanted Shockley to address when he first reported to Bell Labs . . . back in 1937. However, because of its additional structural demands, the device was far more complex than a transistor, and unlike a transistor, it could not be used as an amplifier. Thus, the addressable market was much smaller.

Noyce, Moore, and other new members of the Shockley team were disheartened by the sudden change of focus mandated by their boss, and they tried to persuade him not to pursue this new product. Shockley, however, always sure he knew best, rudely dismissed their suggestion. As Teal and TI had already established a leading position in the silicon transistor market, Shockley felt that his flagship laboratory should not follow in anyone's steps, a mentality common among academic researchers. In order to be successful, Shockley posited, the lab must develop a totally new product, and if they had to address a wholly different market than they'd intended to target, so be it. The young engineers reluctantly gave in to their boss, but tension between Shockley and the group began to foment.

As if his strategic flightiness were not trouble enough, Shockley showed no interest in his company's financial condition whatsoever. As cash became depleted in pursuit of Shockley's new pet project, the company's finances soured, yet Shockley took no actions to right the ship. The four-layer diode's introduction date was repeatedly delayed, but Shockley kept his head buried in the sand, leaving his young team aghast as disaster loomed unheeded on the horizon. Watching the company slide ever closer to ruin, Noyce, Moore, and the other young engineers discussed the matter amongst themselves and eventually decided that they could not remain silent. Convinced that the boss was not fit to run his own company, they decided to go over Shockley's head and talk directly to the financial backer, Beckman, without Shockley's knowledge. There were eight of them: Robert Noyce, Gordon Moore, Julius Blank, Sheldon Roberts, Eugene Kleiner, Victor Grinich, Jay Last, and Jean Hoerni.

During the span of a couple of months, the eight concerned technologists had four meetings with Beckman to discuss the urgent situation at the company. Their proposed plan was to remove Shockley from control of day-to-day operations and hire a professional business manager to take the helm. Beckman grilled them—if an outside business guru was brought in, who would oversee technological and product development and manufacturing? They would, they confidently answered.

Beckman understood the situation, and he was sympathetic to the young engineers. However, Beckman felt he had made not just a professional but a personal commitment to Shockley, and he refused to do as J. P. Morgan had done when he kicked Edison out of GE. Eventually, news of these meetings and this decision trickled back to the rank and file of the company, and when Shockley found out what had happened behind his back, he was both furious and heartbroken. He felt that he had been betrayed by the group of young engineers, allegedly, but famously, branding them "the Traitorous Eight." After this incident, Beckman made some minor organizational changes to try to smooth things out, but the rift proved too great to mend. Though the Eight continued to work at Shockley Semiconductor Laboratory, their hearts were already elsewhere.

In due time, Noyce and the others decided to leave Shockley's company; however, they were not willing to be broken up; they believed that they had a combined, cohesive vision and should all remain in it together. They would leave as a group and, together, would all join a different company. One of the eight, Eugene Kleiner, reached out to his father's Wall Street investment brokerage connections to solicit advice. To his surprise, this investment firm sent two advisors from New York to San Francisco to meet with the group. One of these two was Arthur Rock, a young investor fresh out of Harvard Business School who had a special interest in investing in new technology-based companies. Rock was impressed by the quality of the team, and he felt that, from a pure investment point of view, the agglomerated knowledge and capabilities of these men represented a strong business opportunity. So, instead of directing the Eight to an existing concern, he boldly suggested that they look for investors and set up a company of their own.

THE BIRTH OF VENTURE CAPITAL

Until that fateful meeting, it had never occurred to Noyce and the others to start out into business for themselves. But Rock's suggestion broadened their vision. The Eight became extremely excited by the possibility of being their own bosses; however, they were no blind fools. They knew that the risks of starting a new business—particularly as youths lacking much of any industry track record—were very high, and that conservative and risk-averse banks, including Rock's brokerage firm, would be loath to lend them the money. They decided instead to contact big technology companies and see whether any of them would be interested in backing their venture. Working with Rock, they identified thirty-five firms from the Dow Jones Industrial Average Index whom they believed might be interested in investing in the transistor business. Rock then returned to New York and contacted each of the businesses on their behalf.

In the days that followed, Rock sent out many enquiries. However, the corporations he contacted either turned him down outright or refused to reply altogether. The main problem was that most of the companies that were really interested had already set up their own internal teams to develop transistor technology, and investing in a second, outside venture to engage in similar activities would naturally create conflict and affect employee morale. Thus, they all declined. The rest of the firms simply didn't see how transistors would fit into their existing business models, and they all declined.

Time passed without much progress, and the anxiety level among the members of the group began to rise. Despite their attempted coup, they were still working at Shockley Semiconductor Lab, though after the incident, the work atmosphere had grown depressive and at times outright poisonous. The Eight were all so eager to leave, they began to talk about quitting outright, even if they couldn't remain together and were forced to find jobs on their own. But just when prospects seemed their dimmest, Rock chanced to meet Sherman Fairchild, a New York tycoon whose father had been one of the founders of IBM. Because Sherman had been an only child, he inherited his father's entire estate, becoming IBM's biggest individual shareholder and a very rich man.

In addition to his inherited wealth, Sherman Fairchild owned and operated an aircraft and commercial instrument company on Long Island that specialized in high-performance aerial camera systems. Planes and photography were Fairchild's hobbies, and as a very wealthy man, he had the time and opportunity to indulge them both. Another of his pet interests was flashy, new technology. When Fairchild met Rock at a party and heard his pitch, he became interested in investing in the proposed venture. Soon after, Rock arranged for Fairchild to meet with Noyce and the others in San Francisco. The meeting went well. Fairchild liked what he saw, and he decided to invest $1.5 million to establish the Fairchild Semiconductor Company as a subsidiary of the Fairchild Camera and Instrument Corporation, headquartered in Long Island, New York. For its founding investment, Fairchild Camera and Instrument would own 70 percent of the subsidiary, including buyback rights to the remainder of the ownership stake and the rights to the Fairchild name. Rock's brokerage firm, which made a minority cash investment, would own 20 percent. The remaining 10 percent was to be equally split among the original eight founders; all that was asked of them financially at outset was to invest $500 of their own money as a personal commitment; "skin in the game," so to speak. All parties agreed, and though none of them could have foreseen it, this arrangement led to the birth of the venture capital business model so prevalent today. Rock himself moved from New York to San Francisco, joining the men out west and establishing Silicon Valley's first venture capital fund. Years down the line, when both Intel and Apple Computer came looking for start-up money, Arthur Rock's firm would in each case be the major initial investor.

THE CHANGING OF THE GUARD

By September 1957, everything was ready to go. The Traitorous Eight jointly tendered their resignations from Shockley Labs, and the very next day, they all reported for work at Fairchild Semiconductor. Their newly leased factory space was one empty building only several miles away. At the time, there was no dedicated industry specializing in manufacturing equipment for semiconductor production, so all of the company's equipment had to be specially designed and built from scratch.

Among the eight, Robert Noyce and Jay Last had strong backgrounds in optics, so they bought several used 16mm camera lenses and designed and built three sets of precision light exposure instruments for use in photolithography—light etching—which was a key component in transistor manufacturing. Gordon Moore and Jean Hoerni took charge of designing and constructing several gas-phase diffusion furnaces, and they also constructed the associated quartz tubing plumbing. Moore was a trained glassblower; this ability came in handy and was put to good use. Sheldon Roberts designed and constructed the silicon single-crystal pulling equipment. Vic Grinich, an expert in electronic systems, set up semi-automated equipment to test the performance of transistors as they were produced. As for the last two, both Julius Blank and Eugene Kleiner had previ-

The Fairchild Eight, from left: Gordon Moore, Sheldon Roberts, Eugene Kleiner, Robert Noyce (center), Bob Grinich, Julius Blank, Jean Hoerni, and Jay Last. Wayne Miller/ Magnum Photos

ous manufacturing experience. They were therefore responsible for developing the entire production process flow and integrating all the equipment so it would work together flawlessly. Kleiner was also a skilled machinist, so he designed and crafted on a lathe many of the parts required to fit out the factory floor. Clearly, these eight men were not simple "desk engineers"—they were hands-on builders and able to translate their knowledge into tangible results.

Though it must have been frustrating to be building a transistor production facility from scratch for the second time in as many years, these men were excited and passionate, as they all had real equity in their new company. They were elated to be masters of their own destinies, and there was no more Shockley to get in their way. Without the dictatorial and flighty old man to derail their work, the company's business objectives were clear and unwavering: all their energy was focused on developing the best silicon transistors using diffusion technology. Two months later, they got their first product order: a $500,000 contract from IBM. With this cash injection, Fairchild Semiconductor was able to grow. They expanded their team, hiring many skilled experts from all over the country. Within ten months of first moving into the empty factory building, Fairchild Semiconductor had successfully produced its first batch of transistors and delivered them to IBM. A new company was born.

Later, similar actions mirroring those of the Traitorous Eight recurred countless times in other Silicon Valley companies. If Beckman had only supported what the Eight had proposed and they had stayed, the Shockley Semiconductor Laboratory might have grown into the world's biggest and most successful semiconductor enterprise, and Northern California might never have grown into a hotbed of start-up activity. Likewise, if Rock had not at the critical last minute found Sherman Fairchild to invest in the venture, those eight men might have all gone their separate ways, and Silicon Valley as we know it today might not exist! It is amazing how much fate can seem to hinge on such chance.

In contrast to the success of Fairchild, Shockley Semiconductor Laboratory, its health tenuous from the start, fell into crisis after the Eight departed. Bell Labs tested prototype samples of the four-layer diode Shockley Semiconductor produced and shipped; unfortunately, the new product did not meet the specifications of a central telephone switch. As the Eight had warned, there existed no other major customer or market for this product, so with this failure, Shockley Semiconductor was ruined. Beckman, at last coming to terms with the reality of the situation, quietly sold the business out from under Shockley to a little- known, small-cap company named Clevite.

Looking back, we can appreciate that Shockley indeed possessed a keen eye for recognizing talent. All of his Traitorous Eight were truly top-notch, all-around technologists, and their expertise perfectly complemented one another's. A shame, then, that Shockley himself lacked the personal leadership capabilities to take advantage of the opportunity they presented. Together, they could have driven the company to profitability and the fortune that Shockley was angling for. Instead, the talent went elsewhere, and Shockley's venture failed. It's interesting to note that these Eight were already actively preparing to set up and operate their new company while still at Shockley Semiconductor. Thus, it's certainly possible—even likely—that they brought with them proprietary technologies from Shockley Semiconductor developed while still under that company's employ, perhaps in violation of their contracts. But Beckman never took the Eight to court, although he did consider doing so. Perhaps he held back out of respect for the fact that the eight engineers had approached him numerous times to discuss ways of solving the problems at the company internally before finally being forced to leave. Whatever the cause, Fairchild's success and the concomitant failure of his own company devastated Shockley.

Once his flagship laboratory was sold off to a third party and lost its name, Shockley retreated again to the safe, if less remunerative, haven of academia. He accepted a position on the faculty of Stanford University, relegated to the sideline as Silicon Valley enjoyed its meteoric rise in the years that followed. A broken man, Shockley became somewhat unhinged, ceasing much of his research in physics to focus on the highly controversial field of eugenics—goal-based breeding of humans. He died of prostate cancer at the age of seventy-nine, if not a pariah, at the very least, not the lauded hero he'd hoped to be. Shockley was unquestionably the single most influential person in the history of semiconductors, but his was a flawed genius. Though he was a key contributor to mankind's historical journey of harnessing the tiny electron, his name never reached the heights of fame in the lay press that he'd always hoped for—the most prominent time his name was mentioned in the *Wall Street Journal* came in reporting the fire sale of his namesake firm.

Shockley the Educator

Derek Cheung

As a student at Stanford in the late 1960s and early 1970s, I had the good fortune to take three courses taught by Professor Shockley. My impression of him was that he was a highly intelligent and mellow-tempered elder, one driven by a strong competitive instinct. He never put on airs before his students; he was always extraordinarily serious about teaching and never missed a single class. The textbook we used was his 1950 masterpiece: *Electrons and Holes in Semiconductors*.

In teaching, Shockley stressed the importance of conceptual understanding of a subject, of drawing mental pictures, and of developing quick, order-of-magnitude estimates of the solution. Unlike many other scientists, he cast mathematics as useful primarily in framing a concept and deriving quantitative values, a tool to be employed in detailed engineering design only after a conceptual solution and a rough order of estimation had been reached.

Shockley also had a very vibrant mind that constantly probed for outside-the-box possibilities. He always advocated that, when solving a complex problem, a person should first break it down into a series of simpler problems, and then start to reach a solution by first attacking the simplest block. Shockley frequently made the point that there were many problems for which, for a variety of reasons, there might not be an answer at the time. If a problem could not be solved after a long period of hard work, one should step back and reconsider all the fundamental elements. I remember vividly that during one examination, there was a quantum mechanics problem no one could solve, which left us all perplexed. Later, in discussion, after listening to everybody's views, he let out a great laugh and said, "Has it not occurred to any of you that the boundary conditions I gave contradicted one another?" Slowly, it dawned on us. "There's no solution to this problem since the problem was not defined correctly!" Maybe we had not yet learned to think as broadly as he did, or maybe we were all just too intimidated by such an authoritative man to question the framework he presented. Regardless, this was a valuable lesson to us that we should always question the validity and think through the definition of a problem before tackling it with full force.

Though Shockley admittedly failed in the marketplace, he did not appear to be bitter. In lectures, he never discussed eugenics or the Traitorous Eight, though he did occasionally say a word or two about his bygone days at Bell Labs. He was quite fond of discussing his own critical thinking methodologies, most notably lauding "the will to think" and "creative failure." The will to think was for him exemplified by the two weeks he spent locked up in a hotel room at the end of 1947 to finish his seminal work on the junction transistor. As for creative failure, Shockley loved no example better than Bardeen and Brattain's point-contact transistor—though the device failed in the marketplace, it led to the invention of a superior junction transistor! So creative failures are not to be criticized; with the right mindset, they are simply precursors to more important successes. After all, the fact that Shockley Semiconductor Laboratory went belly-up itself represented a "creative failure"—one that led to the immense success of the Silicon Valley!

16

THE INTEGRATED CIRCUIT AND THE CHIP

In 1957, the Soviet Union successfully launched the Sputnik satellite, shocking and astonishing all of America. In response to this *de facto* challenge, two new government agencies were established in early 1958: NASA, the National Aeronautics and Space Administration, and DARPA, the Defense Advanced Research Projects Agency. These two agencies were charged with quickly regaining and sustaining the United States' position as a global leader in space and military technologies.

Both agencies were provided with a significant budget. This was good news for a still-burgeoning semiconductor industry whose primary clients were involved in the defense industry. Fairchild Semiconductor, its business underway for less than half a year, received its seminal $500,000 contract from IBM to provide advanced silicon transistors for use in computers being custom-built for the military. Not long after, Fairchild Semiconductor was able to obtain procurement and technology development contracts directly from NASA and DARPA, and significant funds started pouring in the company's doors. Very quickly, the company turned profitable, and its transistor technology began to mature. The same thing happened at TI, Motorola, and other semiconductor firms.

With the transistor finally and fully replacing vacuum tubes, it became at last possible to build lightweight and compact electronic guidance and communication systems for space vehicles and rockets. Initially, the transistors were packaged in standard, tiny metal cans as pioneered by Pfann. Even though they were much smaller than vacuum tubes, these cans were still far bigger and heavier than was optimal. One suggested solution was to shrink the cans; a still better one was to eliminate the can altogether.

KILBY AND THE FIRST INTEGRATED CIRCUIT

In 1943, an Austrian researcher in Britain named Paul Eisler invented printed circuit board technology, or PCB. PCB employed tiny chips of unpackaged resistors and capacitors that were directly soldered onto specially designed, metalized connecting patterns printed on a circuit board. U.S. military technical experts realized that the PCB approach held great promise in shrinking the size and weight of entire electronic circuits, while also having the added benefit of improving circuit reliability by reducing the number of separate parts that had to be joined with dangling wires. With funding from the military, an advanced, new semiconductor packaging technology known as "thin-film hybrids" was soon developed based on the original PCB concept. In it, multiple thin layers of ceramic material—an insulator—with specially designed metalized connection patterns were prepared, and these ceramic layers were then precisely stacked together and electrically and vertically interconnected via holes filled with metal pillars. Miniaturized electronic components such as capacitors, resistors, diodes, and unpackaged transistor chips were placed in precise positions on the top of these ceramic stacks. Heat was then carefully applied to solder all these electronic components into the electrical connection patterns on the ceramic substrate, completing a compact, robust, and self-contained circuit.

Jack Kilby. Courtesy of Texas Instruments, Inc.

Employing this thin-film hybrid technology, radar tracking circuits could be made small enough to be mounted inside homing rockets' nose cones. The power and capability of thin film hybrid circuits continued to grow as designers found ways to increase circuit complexity and decrease their physical size. One of the experts among these designers was a young engineer named Jack Kilby.

Jack Kilby was a man of few words, but his thinking pattern was extraordinarily logical and thorough. During World War II, the strapping, young Kilby enlisted in the Army, but at six feet six inches, he was too tall for combat duty, and thus, he was sent to a rear-action camp in Burma. There, his main responsibility was to maintain and repair his unit's radio communications equipment. Due to the extreme humidity of the Asian subcontinent, the Army's standard radio gear broke down frequently. Kilby, an excellent problem solver, made several modifications to the radios' design that greatly enhanced their reliability, making him an invaluable contributor to the health and functionality of his team.

After the war, Kilby finished his education at University of Illinois, where he received his B.S. degree in Electrical Engineering. In 1947, Kilby joined a medium-sized electronics company in Milwaukee known as Centralab, which specialized in designing and building circuits based on thin-film hybrid technology. After half a decade, Kilby grew weary of this work, and as luck would have it, fate provided an answer. In 1952, Centralab sent Kilby to Bell Labs to attend the historical transistor technology transfer classes. Kilby became instantly fascinated by semiconductors, and he hoped to become more involved in this cutting-edge technology. Unfortunately, even though Centralab had licensed the transistor technology, it had neither the plans nor the resources to build up a transistor business, and eventually, Kilby realized he would be forced to move in order to pursue his interests.

In 1958, Kilby was hired by TI to work at its Central Research Lab, a special business unit established by Pat Haggerty to realize his vision of making TI the world leader in transistor technology. Kilby was overwhelmed with excitement at the prospects this job offered. He was a fast learner and an innovative and disciplined thinker, and in short order, new ideas started to take shape in Kilby's mind.

When July of Kilby's first year at TI rolled around, most of the staff left the Lab with their families for summer vacation. But Kilby was a new employee and had not yet accumulated much vacation time. Thus, he stayed, making him for that month one of the lone remaining employees at the whole laboratory. This time and space presented a rare opportunity for Kilby to put his thoughts together without any distraction, and he took advantage of that chance.

With his years of experience at Centralab, Kilby was very familiar with the process of designing circuits using thin-film hybrid technology, and he was also very familiar with the components normally used in those circuits: chip-form transistors, diodes, resistors, capacitors, and inductors. Drawing upon this knowledge with his fresh exposure to the transistor fabrication process, Kilby came to realize that after the diffusion process, germanium or silicon wafers could not only be etched into discrete transistors, but could also be made into other equivalent electronic components. For instance, using only one of the two junctions of a transistor caused it to behave as a diode, and a diode operated in the non-conducting mode behaved as a capacitor. The semiconducting material itself, without a p-n junction in it, was essentially a resistor. The more Kilby thought about it, the clearer it became that all of the necessary components used in a thin-film hybrid circuit could actually be made on a single piece of semiconductor material! As Kilby thought through the fabrication procedure and analyzed the quantitative value of different components that could realistically be fabricated onto a single piece of germanium, he became confident that an entire, useful electronic circuit could be built directly onto a single, tiny piece of germanium.

When the rest of the employees at TI's Central Research Lab returned from their vacations, Kilby approached his superiors and discussed his idea with them, showing them the results of his detailed analysis. They agreed that he was perhaps onto something special, and TI's management provided him with full support to follow his passion and attempt to prove his ideas. Kilby requested and received from lab technicians a small piece of germanium that had already gone through the double diffusion process for transistor fabrication. He then asked them to etch the surface of the sample into a transistor, a diode, a capacitor, and a resistor. Finally, Kilby soldered on gold wires to electrically connect these components into a complete circuit. The wires also provided the added function of being an inductor, completing the full range of component requirements. The circuit Kilby chose to implement was a simple oscillator, the very same design first demonstrated by Edwin Armstrong for generating carrier waves back in 1914 that ushered in the radio era.

As Kilby slowly applied a voltage to the gold wires connected to his extremely crude "integrated circuit" chip, the CRT screen of the measuring instrument that monitored the circuit's output signal began to show some wiggling. With further adjustment, the screen began to display a clear oscillating signal at one million cycles per second, close to his design goal. Kilby had done it. For the first time, an entire circuit had been integrated onto a single piece of semiconductor. On September 12, 1958, the world's first integrated circuit was born!

Immediately, TI management recognized the importance of this invention. They clamped the work in secrecy as they commenced the formulation of a patent application. Shortly after his initial success with germanium, Kilby demonstrated the same concept with the more-versatile silicon, and he began to design and build more complex circuits using still more components etched on to the chip. But through all of this, there remained one unsolvable, fundamental problem: how

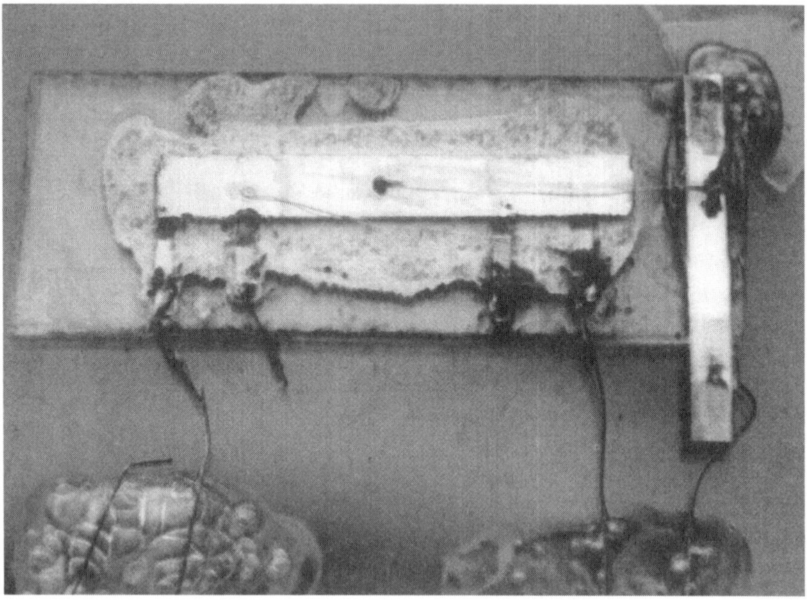

Kilby's first integrated circuit chip. Courtesy of Texas Instruments, Inc.

could he reliably interconnect all the components etched on the one chip? Yes, he could create a functioning circuit all on one chip, but soldering each component with protruding gold wire as he did was just a stopgap solution—it was unwieldy, unreliable, and inelegant, it couldn't be scaled up to build larger circuits, and it would not be mass-producible. Kilby was baffled by this obstacle, and despite his genius, he was not able to determine an acceptable solution. Even as he penned the final patent application for the integrated circuit, the best Kilby could do was vaguely mention the potential to connect the multiple components to one another using thin-film metallization, a concept borrowed from thin-film hybrid technology. But he did not have a specific technical approach to implement the idea that would be compatible with chip fabrication processes. Someone else would have to come up with a specific and practical solution.

HOERNI AND THE PLANAR PROCESS

Jean Hoerni was born in Switzerland, a country whose snowy peaks he loved to climb as a boy. An outstanding student, he was determined from a very young age to grow up and become a research physicist. He received his PhD in physics from the University of Geneva and then went on to Cambridge University to receive a second PhD in theoretical physics. Hoerni moved to Cal Tech in 1954 to do post-doctoral research; while there, he met and impressed visiting professor William Shockley, who at the time was still searching for the next step in his career after leaving Bell Labs.

Jean Hoerni. Wayne Miller/ Magnum Photos

In 1955, when Shockley was recruiting for his new startup, he thought of Hoerni and made a special trip to Cal Tech to recruit him. Hoerni had never thought about working in industry, but Shockley's sincere gesture and generous offer were hard to refuse. Hoerni decided to give it a try and joined Shockley Semiconductor Laboratory. However, Hoerni, like so many others, chafed under Shockley's brusque and uninformed managerial style, and he became one of the Traitorous Eight who left to found Fairchild Semiconductor.

By the time Kilby joined TI, Hoerni and his colleagues at Fairchild were already hard at work developing diffusion technology to mass-producing silicon transistors. The performance of Fairchild's transistors was excellent, and their popularity with their customers was reflected in their rapid increases in sales volume. However, in spite of their commercial success, production yield for transistors remained low, and some crucial reliability issues remained unresolved.

Traditionally, after the impurity diffusion process to form p-n junctions was completed, the wafers were etched using photolithographic techniques to form isolated, independent transistors. The etched profile of each completed transistor then looked like a miniaturized desert mesa: a flat, broad peak with steep slopes. This led to the moniker "mesa transistors." The p-n junctions exposed on the mesas' steep slopes were very sensitive, finicky, and difficult to protect, which led to the yield and reliability problem with which all transistor manufacturing companies struggled.

The basic technology that led to the solution of this critical problem originated, perhaps unsurprisingly, from Bell Labs. In 1954, a team of scientists led by Carl Frosch was conducting research to further refine gas phase impurity diffusion techniques in silicon. By accident, they discovered that if water vapor were added to the gas flow during the high-temperature diffusion process, it would minimize damage to the silicon surface. Normally, as the diffusion process proceeded, the surface of the silicon would become damaged and pitted by the gas. The presence of steam, however, oxidized the silicon surface, producing a conformal, thin layer of glassy silicon dioxide (SiO_2). This silicon dioxide layer, continuous, dense, chemically stable, and electrically insulating, would serve as a perfect shield, protecting the silicon underneath from the environment. This auto-insulation is a unique property to Si and this "thermally grown" silicon dioxide; no other semiconductors, including germanium, exhibit the same behavior. In 1956, Frosch's team published this important finding, but the paper did not garner much attention from the burgeoning industry.

At this time, the upstart Fairchild Semiconductor was hard at work filling high volumes of purchase orders from customers. As in any typical, rapidly growing start-up environment, everyone was busy all the time—developing novel products, solving immediate problems, and responding to customer demands. At the time, Hoerni was responsible for the production of one type of the transistor that was plagued with a low yield problem, so he was under constant pressure. However, unlike many other engineers and managers who did not have the time and energy to think or read about new and basic technologies, Hoerni, ever the scholar, always found time to keep current of new technical advances, especially for

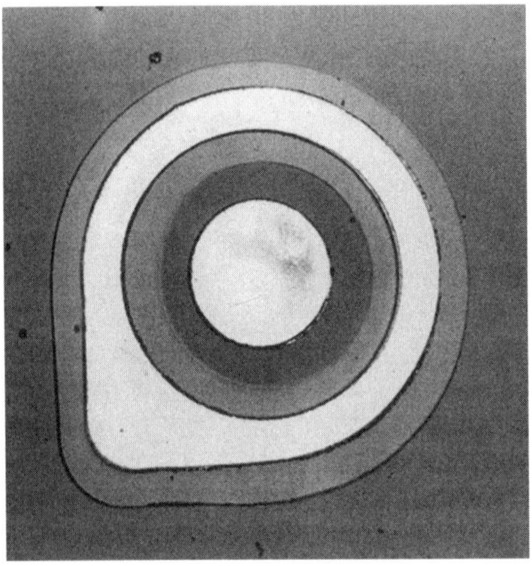

Hoerni's first planar transistor. Courtesy of Fairchild Semiconductors

semiconductor research coming out of Bell Labs. He was aware of Frosch's work on the fundamental properties of silicon and silicon dioxide, and in early 1957, it occurred to him that there might be a way to apply this knowledge to creating a new and better fabrication process for silicon transistors. Hoerni spent the better part of a year trying to conceive of methods to leverage Frosch's ideas—exhibiting, as Shockley would have said, "the will to think"—and by late 1957, he had completed an initial design of a completely new procedure to fabricate transistors.

Hoerni's new method was to repeatedly employ a combination of photolithography, etching, diffusion/oxidation, and thin-film metallization on a silicon wafer. This approach took full advantage of using the thermally grown silicon dioxide layer, formed naturally during the diffusion process, to effectively protect and seal off the sensitive silicon surfaces—especially the exposed p-n junctions—from the environment, like a cocoon providing protection to delicate silkworm larvae inside. A key advantage of Hoerni's new process was that all transistors were formed on the same flat plane, with no etched mesas with their sharp edges and steep sides. Thus, Hoerni named the new manufacturing method the "planar process."

At outset, nobody at Fairchild, even the visionaries Noyce and Moore, realized the importance of the planar process. They were all too busy furiously constructing mesa transistors to meet delivery schedule; there was no excess capacity among people or equipment to expend on experimentation. As a result, Hoerni's new fabrication idea sat on the shelf, undeveloped, for over a year. Finally, in January 1959, Hoerni was able to bootleg—from the bottom up—the entire process and fabricate his first planar transistor. Upon testing, to everyone's surprise but his own, Hoerni found that transistors fabricated via the planar process far outperformed mesa transistors in uniformity, reliability, and yield, all elements critical to their successful mass production and reliability in applications. With Hoerni's planar process, Fairchild had a winning technology, truly a major breakthrough in transistor design and fabrication. As Fairchild started to apply for patent protection of the process, they also made a courageous business decision: from that point forward, all Fairchild transistors would be produced via the new planar process.

NOYCE AND THE CHIP

The successful development and implementation of the planar process was a major event at Fairchild, one with far-reaching impacts upon the entire industry. During a meeting regarding the patent application, Fairchild Semiconductor's patent attorney asked the company's technical staff if there were other, related ideas that ought to be included in the invention claims. This stimulated Noyce's will to think. As he considered the flatness and unity of the planar process, he conceived that it might be possible to use thin-film metallization over multiple insulating silicon dioxide levels to interconnect all the various components of a complete circuit onto one semiconductor chip. Though Noyce had no specific knowledge of the work Kilby was doing at TI, he was clearly thinking along the same lines. Noyce's breakthrough came after Kilby's initial demonstration, but thanks to Noyce's understanding of Hoerni's work, he realized that through a simple extension of the planar process, a *monolithic* integrated circuit—one with no extruding wires for interconnections—could be designed and produced. The problem Kilby had encountered and suffered in silence was solved, independently, by Noyce and the Fairchild team.

After extensive analysis and experimentation, Noyce designed a simple flip-flop logic circuit based on his new concept. In 1961, one of the Traitorous Eight, Jay Last, successfully fabricated the new integrated circuit using the planar process. This integrated circuit consisted of two transistors, several resistors, and a capacitor. All components were fully integrated and interconnected on a single silicon chip and there were no dangling wires. To distinguish between this approach and Kilby's, the Noyce version later became known as the "Monolithic Integrated Circuit," or more commonly, the *chip*.

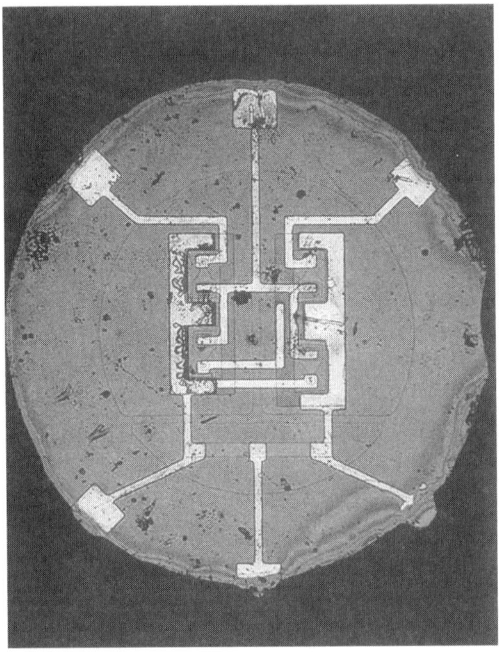

First monolithic integrated circuit, or chip, designed by Robert Noyce and built by Jay Last. Courtesy of Fairchild Semiconductors

The chip functioned flawlessly. More importantly, it was born production-ready: it was practical, reliable, scalable, and mass producible. This was a historic achievement, and it marked the beginning of the modern electronic era. With the chip, mankind could now precisely control the complex flow of electrons in a functional circuit built on a tiny piece of silicon. Based on the fundamental design concept and fabrication technique, chip technology would grow dramatically over the next fifty years. The first flip-flop chip, made in 1961, contained only two transistors; a chip of comparable size today can contain more than five billion transistors. The situation is like comparing today's A-380 and Boeing 787 with the Wright brothers' first airplane in 1903—the basic design and the aerodynamic principles are the same, but the complexities involved are as different as night and day.

Though the chip was indeed magnificent, the legal battles it spawned were assuredly not. Although Kilby's patent application was filed earlier than Noyce's, Noyce's was nonetheless granted first. TI appealed, and the ruling was reversed; a subsequent re-appeal by Fairchild led to yet another reversal. This string repeated itself until it became clear that the two sides would have to find a way to compromise. TI wanted to own an exclusive patent for the integrated circuit, but it was clear that they could not feasibly manufacture the device without Fairchild's planar process. Similarly, Noyce and the other young technologists at Fairchild knew that even though they were the first to demonstrate a practical integrated circuit, they had not been first in aggregating various components onto one chip. Gradually, both sides came to the realization that the best result was for the two companies to compromise. In 1966, Fairchild and TI reached a commercial agreement. They decided that they would simply cross-license their patents with each other; however, other companies who wanted to use their technology would have to negotiate licensing agreements with TI and Fairchild separately.

Though corporate leadership of both companies acquiesced to this deal, lawyers on both sides nonetheless continued to drag their battle all the way to the Supreme Court. In 1971, the Court finally declared that Noyce was the true inventor of the integrated circuit, but the results of this judgment did not carry any real significance. An accord had been reached five years earlier, and both TI and Fairchild collected hundreds of millions of dollars in licensing fees. This proved to be an important source of profits for both companies during this early stage of the chip industry.

Despite the Supreme Court's decision (and the fact that they worked completely independently of one another), Kilby and Noyce remain to this day jointly credited with inventing the integrated circuit. But without Hoerni's planar process, Noyce would never have been able to invent the chip in the first place—nor would either of them have been able to do so without cleverly incorporating an invention by an engineer named Kurt Lehovec. Lehovec introduced an elegant method of using diffusion technology to electrically isolate various components on a planar chip, and this technique was crucial for Noyce's chip to function. Thus, all four men deserved credit for the roles they played in developing the chip.

Hoerni's planar process was the cornerstone of the chip technology; however, he was not well known outside of the early semiconductor industry. In 1961, shortly after his invention of the planar process, Hoerni left Fairchild. He became a serial entrepreneur and subsequently created three successful semiconductor companies. He continued to pursue his lifelong passion of mountain climbing and would spend every summer climbing the Kunlun mountain range in Kashmir and Afghanistan. Upon his death in 1997, he donated a significant portion of his wealth to the Central Asia Institute to provide education to women and children in that part of the world, as well as to help would-be mountain climbers to realize their dreams.

In 2000, Kilby accepted the Nobel Prize in Physics for his invention of the integrated circuit. By then, both Noyce and Hoerni had died, and only living candidates are considered for Nobel Prizes. In accepting this honor, Kilby was in a sense celebrating Noyce, as well as all those others who had made important contributions to the development of chip technology. Kilby always viewed himself more as a problem-solving engineer than a knowledge-creating scientist. Despite the corporate legal battles between his and Noyce's employers, he accepted the Nobel Prize with a spirit of great humility and comradeship, and when he died in 2005 at age eighty-one, the entire electronics industry mourned the loss.

As for Noyce, his legend only grows from here.

Before ending the discussion on the chip, it should be pointed out that, as with many other revolutionary technologies, enthusiasm and support for the chip was not universal, especially in the early years. The leading skeptic was Jack Morton, early management champion of the transistor and eventually successor to Mervin Kelly as the Director of Research at Bell Labs. Based on his firsthand experience in leading the effort to mass produce transistors, Morton pointed out that the low yield of individual transistors would forever limit the potential success of practical, large-scale integrated circuits. It was simple statistics, he claimed: the more transistors on a chip, the lower the chip yield, and the higher the likelihood of chip failure. At the time, transistors were fabricated with the mesa process, and both yield and reliability were extremely low. Widely viewed as an authority in the field, Morton and his "tyranny of numbers" concept appeared to be convincing and gained many followers. But the reality was that Morton had become straightjacketed by his personal experience. Noyce and his colleagues fearlessly advanced chip technology with the help of breakthrough technologies such as the planar process, which were not accounted for in Morton's incremental line of thinking. Reflecting on these events many years later led Noyce to coin a famous quote, which he directed to young people everywhere: "Don't be encumbered by history; go off and do something wonderful!"

FAIRCHILD AND THE SILICON VALLEY PHENOMENON

Thanks to Noyce's leadership, the technical superiority of the planar process, and the success of the monolithic integrated circuit, Fairchild Semiconductor had by the dawn of the 1960s established itself as a leader in the semiconductor industry, with a strong reputation for innovation that attracted much top talent from all over the world. Barely two years after Fairchild Semiconductor had been launched, its parent corporation, Fairchild Camera and Instrument Company, took a close look at the business and became convinced that the outlook of the new organization was very bright. It decided to exercise its rights to buy back the 30 percent equity owned by Arthur Rock's company and the original eight founders. After negotiations, the parent firm agreed to pay $6 million for the outstanding equity. This yielded $250,000 for each of the Traitorous Eight, a handsome return on their $500 initial investments! Though this was quite a windfall, the founders nonetheless viewed this development with mixed feelings. On one hand, they felt a sense of satisfaction in receiving such a reward for their success. But on the other hand, they no longer owned any piece of the company that they had toiled so hard to build. Now, they were mere employees, working only for a salary. This realization hit them hard, and the passion and dedication they had previously devoted to their work diminished. Their entrepreneurial spirit, too, flared up in protest—and now, with fortunes of a quarter-million dollars each, they were economically independent. If they wanted to leave Fairchild, there remained nothing to hold them back.

After the parent company bought back its equity, the New York headquarters appointed Noyce as General Manager of the new semiconductor division. This subtly changed the relationship among the original eight founders: now, seven had to answer to one man; one among them had become their boss. To some, this was untenable, and in early 1961, the newly wealthy Hoerni, Last, Roberts, and Kleiner left Fairchild together to establish a new semiconductor division at a company called Teledyne. Again, Arthur Rock acted as the go-between. Interestingly, Kleiner didn't stay at Teledyne long: following the model he'd learned from Rock, he soon left to form his own venture capital investment group. This firm, Kleiner-Perkins, eventually evolved into one of the most important venture capital groups in Silicon Valley, playing a major role in seeding revolutionary technology companies like Amazon, Google, and Genentech.

As four of the original founders departed, it started to become clear to many would-be entrepreneurial employees at Fairchild that there was no future in staying at the company. Fairchild Camera and Instrument Company, the corporate parent, was based on the other side of the continent in Long Island, New York, and it was an extremely conservative and traditional firm. Their corporate culture and operating style constantly clashed with those of the semiconductor operation in Silicon Valley. Even though Sherman Fairchild had been heavily involved in starting the semiconductor business, he did not much involve himself in its subsequent operation, relying instead on his lieutenants to oversee its administration. But these business managers from the east coast knew nothing about the semiconductor business and cared even less; their only interest lay in looking for short-term financial gain.

As Fairchild Semiconductor generated cash flow, its headquarters in New York siphoned away almost all of its profits, leaving behind only a small amount for reinvestment. This shortage of capital significantly and artificially slowed Fairchild Semiconductor's growth at a time when the market was expanding rapidly and prospects for successes abounded. Many golden opportunities slipped away as the semiconductor division's proposed projects and funding plans, all of which needed approval from New York, were met with sluggish and at times irrational responses. Noyce and his team grew ever-more frustrated in their dealings with the inept headquarters. Bitter conflicts and frictions kept on mounting, which began to seriously damage employee morale. Meanwhile, the venture capital system pioneered by Rock was just taking shape in Northern California, and suddenly, for a qualified entrepreneur with a strong business plan and a winning technology, it was not difficult to exit one's job, raise start-up funds, and start a company all one's own.

The seismic shift in business philosophy initiated by Arthur Rock and personified by the Traitorous Eight led to an up-swell of entrepreneurialism. In 1959, just one year after Fairchild Semiconductor was created, its first business manager abdicated with a small group of engineers to start yet another new transistor company, Rheem Semiconductor. What the Traitorous Eight had done to Shockley was played back on them, and what's more, it was no isolated incident! Over the next ten years, this wave of spinoffs never stopped, in fact growing even stronger all the while. Again and again, entrepreneurial employees would leave the company and start up their own businesses, bringing their newly developed technologies and product ideas with them. Other than the first such spin-off, Rheem Semiconductor, Fairchild management chose to not litigate against any of these corporate offspring, and this turning of a blind eye served as tacit encouragement for people to go off on their own. Perhaps the leadership was planning to pack up and leave themselves at the right moment all along, or perhaps they simply felt they owed greater allegiance to their friends and former colleagues than to the remote and faceless Fairchild headquarters staff. Maybe it was just that they appreciated the fact that Beckman had never tried to sue them, and they collegially paid this genteel treatment forward. Whatever the case, spinoff companies began to dot the marketplace like spores cast off into a meadow, and their numbers simply multiplied.

By 1961, Arthur Rock had moved permanently from Wall Street to Northern California, leapfrogging from the role of a junior associate in an east-coast investment banking establishment to that of the leading figure of a nascent venture capital industry. As the relationship between Fairchild Semiconductor and its parent company worsened, Rock counseled Noyce and Moore, as he had in the past, to simply exit and set up their own business. After all, they'd done it once before, and with their new reputations as the leaders of the semiconductor industry, raising the necessary capital would be no problem at all. Noyce and Moore remained reluctant, though—at least, at first. Eventually, they came to realize that, as at Shockley Labs before, the situation at Fairchild had become so dire that it was simply untenable.

In 1968, the semiconductor industry was on the verge of exploding into an enormous new market: computer memory chips. At the same time, the top management at Fairchild Camera and Instrument Corporation was going through a state of flux. The semiconductor division was responsible for 110 percent of the entire Fairchild group's profits—the net profit of all of the other groups was negative, and only Noyce's semiconductor business was in the black. By all rights, this should have won Noyce the presidency of the company. But it didn't. The east-coast board of directors saw Noyce as an outsider and did not trust him, and instead, they snubbed him. Disgusted and unwilling to take any more, Noyce and Moore finally made up their minds to leave Fairchild Semiconductor, a company in which they had invested nine years of their lives to shape, build, and lead. The emerging market of computer memory chips provided an excellent focus for their new business, since there were no incumbent companies in this business space and no direct business conflicts with Fairchild.

Noyce telephoned Rock and informed him of his decision. That phone call was placed in the morning; by the end of the afternoon, Rock had already raised $2.5 million in convertible debentures. Noyce, Moore, and Rock would each contribute more to bring that total initial investment to $3.0 million. They raised this money exceedingly quickly—the business plan for the new company was written by Rock himself and consisted of only one typed page! But it was no matter. A few weeks later, Noyce and Moore left Fairchild for good, setting up their own company not too far away. In fact, the founders used what they proposed to make—integrated electronics products—to derive the new company's name: Intel. (They had thought about emulating Hewlett-Packard, and naming their company Moore-Noyce. But it sounded like "More Noise," which was certainly not desirable in the electronics industry!)

In the last major wave of spinoffs from 1967 to 1969, many important companies were born out of the crumbling foundation of Fairchild Semiconductor. In 1967, Charlie Sporck, vice president in charge of production, left to rejuvenate National Semiconductor Company. Jerry Sanders, vice president of sales, created AMD in 1969. Noyce and Moore started Intel. And rounding out the entrepreneurial tale of the remaining Traitorous Eight, Julius Blank, who initially stayed at Fairchild after Noyce and Moore left, later started his own company, Xicor. The final member, Victor Grinich, went into academia, spending many years lecturing at Stanford University and UC-Berkeley.

Under the direction of Noyce and Moore, Intel quickly grew into its role as the new leader of the semiconductor industry, and Fairchild suffered a critical blow. Interestingly, though the original Fairchild would eventually dissolve into the mist of history, its impact and footprint remain strong today: over 400 companies in Silicon Valley today directly or indirectly owe their origin to Fairchild. One company giving birth to so many fertile corporate offspring is a phenomenon wholly unique in the history of industrial development. The family-tree phenomenon was due largely to the combination of that company's incredible talent pool, coupled with its parent company's dismal management failures. People who came out of Fairchild possessed key technical know-how and creative energy as well as the drive to strike out on their own. When workers saw the independent success of one ex-Fairchild employee after another, it ignited their ambition from within, unleashing a big bang of unstoppable force that led directly to the birth of the Silicon Valley success story. It is interesting to note that the success of the Silicon Valley has been a cultural and ground level swell; the government did not have a special policy in place and played no deliberate role in it.

Of course, Fairchild Semiconductor was far from the only cradle of chip technology. TI, Motorola, IBM, RCA, and others were all important contributors, too. But management in those organizations was far more disciplined, and the culture was more conservative. So the impact of spin-off activities from these companies was relatively minor.

Working at Fairchild in the Late 1960s

Derek Cheung

From 1969 through 1972, I was fortunate to have the opportunity to work as a junior engineer in the R&D lab and later in the transistor production line at Fairchild Semiconductor. It was my first job after graduating from university, and looking back, I'm sure I was totally inexperienced, both in working with technology and working with people. I was too green to understand or appreciate the significance of all the shakeups in the organization; I only remember that when I first reported to work in early 1969, Noyce had been gone for just about half a year.

Faced with a leadership vacuum, Fairchild Camera and Instrument paid dearly to lure away Les Hogan, formerly the General Manager at Motorola's semiconductor business unit, to replace Noyce. Hogan brought more than fifty people with him from Motorola, nicknamed Hogan's Heroes, to take over Fairchild. This sudden influx of powerful outsiders created a serious cultural and business philosophy clash with the existing staff, pushing the already-depressed staff morale at Fairchild even lower. Not long afterwards, there was another wave of exodus of top talents to Intel and other startups. Still, the company was brimming with incredible technical talent and those top minds were hard at work pursuing a slew of cutting-edge technologies with outstanding market potential.

I was hired into a special program aimed at developing young, high potential employees and, as a part of the training, I was exposed to a host of groundbreaking technologies and projects. The first project I participated on was designing a 256-bit memory chip. Later, I took part in researching charge-coupled devices, or CCDs, for memory applications. Still another project team on which I was staffed pulled single-crystal gallium phosphide, a new semiconductor material, to fabricate green LEDs. In time, I was even assigned to work directly on the production line, responsible for operating diffusion furnaces for the production of high-voltage power transistors used in television sets.

The depth and breadth of experience I gleaned at Fairchild was invaluable. In addition to the on-the-job training it offered, Fairchild also conducted an intense internal technology training class for its technical staff, complete with formal homework and examinations. Every subtopic in the course was taught by one of the company's specialists, all of whom were at the top of their fields. The teaching material in the class was a combination of theoretical and practical information, and much of the newest knowledge was freshly developed in Fairchild's own laboratories. The original curriculum had been created by Bob Noyce and Andy Grove when they were still at Fairchild, and it had a huge impact in terms of creating and shaping a large number of practically trained engineers in the Silicon Valley. Though Fairchild Semiconductor as a business may not have been managed properly, its technical training program was second to none. At the end of the training, every employee could successfully design a transistor or simple chip to meet a given product performance specification, and every employee could detail the exact fabrication procedures needed to produce such a product.

Part of the value of being at Fairchild was the breadth of projects on which the firm worked. I can still recall sitting down for lunch at numerous times in the company cafeteria amongst engineers from all different departments, eager to chat about technical advances over their sandwiches. Often, the topics discussed ranged across a wide swath of hot, new technologies. Just as often, though, people talked about how so-and-so had gone off to Intel or had left to found a new company. Occasionally, disgruntled workers would complain about how the stiffs in Long Island had screwed up yet another good opportunity, or how production and sales people were stupid not to appreciate the great technologies coming out of the lab. The near-constant refrain was, "If only we could do it ourselves!"

Finally, one day, Frederico Faggin, an engineer from Italy whose office was next to mine, decided to jump ship to Intel. Faggin was incredibly intelligent, a workaholic, and well-respected. I still remembered joining everyone in a farewell lunch for him. While still at Fairchild, Faggin was a key member of a team that developed "silicon gate" technology, which used highly conductive polycrystalline silicon film to replace aluminum in forming gate electrodes. The new silicon gate technology allowed chips to grow still smaller and run faster and more reliably than ever before, all while using less power. I had no doubt that Faggin's inventiveness and knowledge, for which he was known throughout Fairchild, would prove very useful to him at Intel as well, and they were.

A few months after Faggin left, a colleague who was in charge of developing photomasks—negative photographic plates used in photolithography to define fine line patterns on chips—told me that he, too, was leaving. "Do you want to come with me?" he offered. When I asked him which company he was joining, he confidently replied, "I'm going to start my own! We'll specialize in making photomasks, too—customized ones, and we'll be quicker, cheaper, and better than this place! Are you in?" Well, as far as I was concerned, making photomasks was only one small tenet of the entire chip fabrication process, so how could an independent business stand on that facet alone? I declined. Of course, in time, he did succeed, and when he took me out for lunch not long after in his new, cherry-red Porsche 914, I was really envious. The confident and satisfied expression on his face as he dropped me off afterwards instilled a very clear message within me: the energy and creativity in a motivated, courageous person was boundless. You did not always have to follow rules laid down by others; follow your heart and just do it your way!

Although top engineers kept leaving Fairchild Semiconductor, outstanding new talent never stopped joining, either. Fairchild drew the best minds from all over the world, and even though a large number of them would eventually leave to start their own businesses after their "apprenticeships" there, just being present at the epicenter of this historic maelstrom of creativity and entrepreneurialism made me feel lucky to live through such an experience.

17

CHIP TECHNOLOGY BLOSSOMS

THE EARLY MARKET FOR CHIPS

When Noyce successfully demonstrated the first flip-flop logic circuit on a single piece of silicon in 1961, it laid the foundation for the modern information era. Of course, no one could have fully known that at the time. As an emerging technology, the price of these initial chips was relatively high. More importantly, as they were new, there was a shortfall of reliability data. With a high price-point barrier and no guarantee of operational reliability, engineers remained reluctant to design these chips into any of their products. Only one factor spurred demand for and production of chips in those early years: the threat of global war.

The world in the early 1960s was caught in the grip of Cold War. As tensions rose ever higher, the United States and the Soviet Union became engaged in intense competition over the buildup of military and space capabilities. The U.S. military was developing new intercontinental ballistic missile (ICBM) systems, such as the land-based Minuteman missile and the submarine-launched Polaris missile, and NASA was fully engaged in its Apollo moon landing program. Both these space and military applications placed high value on size, weight, and reliability of onboard electronics, whereas sticker prices were secondary considerations. Thus, it was almost exclusively the defense and the aerospace industries that supported the demand for chip products through the first half of the 1960s, much like military needs of the 1950s supported the transistor in its early years.

As it became clear that integrated circuits could indeed replace large numbers of discrete transistors, the military made deep investments in bettering chips' reliability through improvements in fabrication and packaging technology. These cash infusions kept chip-making units afloat and allowed companies to solve many initial problems while accumulating valuable production experience and a reliable database. The military got what they wanted, and every chip manufacturer—Fairchild, TI, Motorola, and others—benefited greatly from this process.

By 1964, chips had become accepted as a better and more reliable replacement for discrete transistors, and hot on the heels of military acceptance came the early adopters from private industry. The first to come calling were, once again, IBM and the other computer firms. The computer industry started to adopt simple chips with logic circuits in a new generation of high performance computers, replacing large numbers of discrete transistors that themselves had previously replaced vacuum tubes. Thanks to this increasing demand, a large commercial chip market began to emerge. From this point forward, chips grew ever more sophisticated, and the number of transistors on each chip began to increase at an exponential rate.

In those early days, all transistors used in making chips were based on Shockley's junction transistor design and fabricated with Hoerni's planar process. Although the performance of the transistors themselves was good, the manufacturing technology used to produce these transistors was relatively complicated, and power consumption of each device remained high. Nagging remnants of production issues kept costs artificially inflated, and each individual transistor's high power consumption led to an even bigger problem: more power consumed meant more heat generated, and when many transistors were packed densely together, chips' performance and reliability suffered due to overheating. To really push chip technology forward, a simpler transistor design with lower power consumption was required.

In 1958, M. M. Atalla and his team at Bell Labs were exploring the potential application of the electronic properties of the interface between silicon and silicon dioxide film. Up until that point, the oxide was used primarily as a passive protective layer; but Atalla wondered if maybe they could be a part of the active transistor as well? Atalla's team discovered that by applying the proper voltage to a metal gate electrode deposited on top of the protective silicon dioxide layer, they could create an electrically conductive "channel" at the silicon-silicon dioxide interface under the gate. The electrical conductivity of the channel could be modulated by the voltage applied to the gate electrode. Though this idea was somewhat analogous to the concept of the electric field-effect transistor first proposed by Shockley in 1938, the difference was that the sensitivity and ease of modulating the conductivity of the channels were far superior to the bulk material. In 1960, based on this principle, Atalla successfully fabricated the first metal-oxide-semiconductor field-effect transistor, or MOS. The MOS transistor truly ushered in the new era of chip technology.

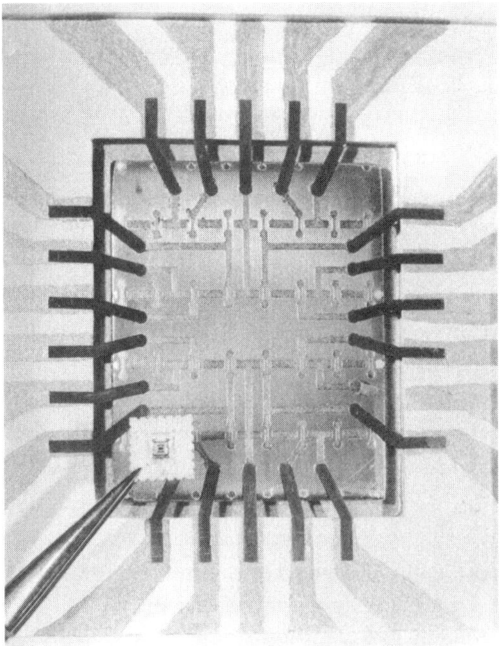

First MOS chip from RCA. Courtesy of Hagley Museum and Library

In a MOS transistor the conductive channel can be either formed by electrons, thus making it n-channel, or holes, making it p-channel. MOS transistors have two important advantages over junction transistors: first, the fabrication process is much simpler and more amendable to scaling, which is to say that MOS transistors can be readily shrunk down in size. The second is that power consumption is much lower in MOS transistors than in junction transistors and less power consumption means less heat generated. Less heat, in turn, means more transistors can be squeezed into a smaller area while maintaining good reliability. These major advantages made the MOS transistor ideal for large-scale chips.

In 1962, RCA Labs successfully fabricated the first chip using the MOS technology, and over the next few years, MOS chips completely replaced ones made with junction transistors. In 1963, Frank Wanlass and C. T. Sah, research engineers at Fairchild Semiconductor, realized that if they could combine n-channel and p-channel MOS transistors onto the same chip, their power consumption could be reduced even further. This dual-channel technology was named Complementary MOS, or CMOS for short, and in 1967, RCA Laboratories successfully demonstrated the world's first CMOS chip. From that point on, CMOS became the mainstream transistor technology

MOORE'S LAW

By the mid-1960s, most of the chip products sold by Fairchild and TI were logic circuits used in computers. Each chip included between ten and one hundred transistors. As computers grew faster and more powerful, market needs and economic drivers began to call for still more and more logic functions and capabilities to be aggregated into a single chip, but without significantly increasing their cost. This created a quandary: how could the number of transistors on a chip be increased without driving up chip size and cost? The solution was to shrink the size of each transistor and the footprint of the metallization patterns that connected them, so more transistor circuits could fit into a fixed chip area with minimum added production cost.

Thanks to the scaling properties of MOS technology, more and more transistors could be bunched ever more tightly on a single chip. In addition, shrinking transistor sizes had other important ancillary benefits, such as reduced power consumption for each transistor and increased operating speed. As a result, ever more complex, faster, and higher-performance chips could be designed and produced, all with a negligible increase in chip production costs. Starting from the mid-1960s, the major focus of chip technology development was the continuous shrinking of transistor size. The number of transistors contained in each chip blossomed from a handful to several hundred.

In 1965, Fairchild Semiconductor's Gordon Moore looked back over the short history of his industry and noticed that between 1962 and 1965, the number of transistors on each chip product doubled every twelve months. Inspired by the data, Moore published a paper in which he famously forecasted that this trend could be continued for many years to come. His prescience became institutionalized across the industry as Moore's Law, and though it was later amended to forecast that the number of transistors on a chip would double approximately every eighteen months rather than every twelve, his general observation remained surprisingly accurate. His perception, of course, was not really a "law" in the scientific sense, but rather a rational and farsighted technology roadmap for the still-young chip industry. In a later interview, Moore self-effacingly admitted that at the time the paper was presented, he felt more like a salesperson making a pitch than a true scientist. He wanted to convince his customers that chip-based products from Fairchild Semiconductor would continue to evolve with time, becoming ever more advanced as time marched on. So the customers should keep their business with Fairchild.

Regardless of what his motives were, Moore's projections held up remarkably well. In 1961, Noyce's first flip-flop chip employed two transistors. Today (2013), a similarly sized chip, such as Intel's 62-core Xeon Phi microprocessor, contains five billion transistors! In the span of about fifty years, the area density of transistors on a chip has increased well over one billion times, without a significant increase in the basic fabrication cost of a chip! There is no other

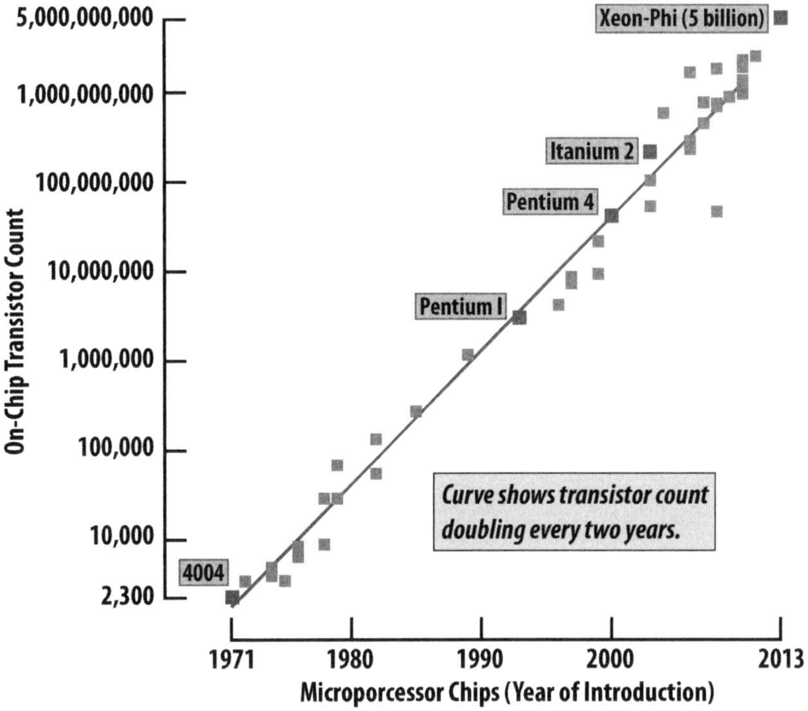

Moore's Law for microprocessors. It is a logarithmic plot of the number of transistors on a microprocessor chip versus the year the product is introduced. The transistor count doubles approximately every two years. Derek Cheung

industry in all of human history that can boast such astounding and sustained advancement. In the ladder-climbing effort of Moore's Law, every time a chip fabrication, design, or construction methodology has neared its natural physical limit for further scaling, a novel technology that may have seemed supernatural a few years earlier was introduced, allowing chips to break through progressively larger and larger technical barriers. When the resolution of metalized lines and transistor dimensions neared the resolution limit of conventional photolithography, new "deep UV" systems with liquid-immersion techniques were developed to continue the march toward ever smaller, nano-scale devices. When gas phase diffusion technology reached its limit of precision control of impurity atoms in silicon, it was replaced by "ion implantation," in which charged impurity ions were shot into silicon at precisely controlled doses and depth profiles. When the accuracy of using liquid acid to etch silicon and silicon dioxide had gone as far as it could, a new plasma-assisted dry etching technology emerged.

The invention and adoption of each of these new technologies marked remarkable advancements in their own right, but still more amazing was that all these technologies were integrated seamlessly to enable the continuous realization of Moore's Law. It is not an overstatement to say that such technological progress has been one of the most fundamental driving forces of the productivity improvement and economic growth of our society in the last fifty years. Notice how whenever you buy electronic products, it always seems that within several months, an improved version of the same product will somehow be available and selling for less money? This is an exemplification of Moore's Law at work.

MEMORY CHIPS

For the past half-century, Moore's Law has served as a flag in the sand, a goal that drives progress in the chip-making industry. As of 2013, many chip products can hold several billion transistors, and the dimensions of some components have shrunk to twenty nanometers—that's twenty *billionths* of a meter! But as wonderful as these advances are, they don't answer the basic question: why do we need so many transistors on a chip?

Initially, the answer to this question was the ratio of cost to performance demanded by the computer industry. This ratio inevitably improves as more and more logic circuits are aggregated onto a chip. The computer industry and the chip industry enjoyed a state of symbiosis: all the advantages gained by chips would trickle downstream into the enhanced performance of computing products, and in return, the expansion of the computer industry brought ever-growing business opportunities for chips. In fact, farsighted and well-heeled computer companies such as IBM began investing into their own internal semiconductor research; similarly chip companies, such as Intel, have acquired tremendous in-house knowledge on computer architecture.

In addition to their initial use as logic processors, chips also proved excellent at performing high-speed memory storage and retrieval tasks, such as interacting with computers' central processor units. In the 1950s and 1960s, computers relied on magnetic core memory, in which the magnetic orientation of tiny magnetic cores could be set one way or another via the pulse of an electric current. This memory solution was pioneered by An Wang, a Harvard-trained engineer who had worked on the Mark IV computer, a digital descendant of the Mark III electromechanical computer. An entrepreneur, Wang took full advantage of his invention and set up Wang Labs as a major supplier of core computer memory. Though he did well, the limitations of magnetic core memory were many, the most serious of which were cost, physical size, and capacity limitations. A new and better solution was needed, and the silicon chip held great potential.

In 1967, IBM engineer Robert Dennard designed a memory circuit comprised of a single transistor and a tiny capacitor. The capacitor had two states: it could either be charged or not. These states, like the directional magnetization of a magnetic core, could be represented as a "0" or "1," and thus, Dennard's memory circuit could serve as a replacement of magnetic core memory in many applications. This memory circuit, which Dennard called DRAM, for Dynamic Random Access Memory, proved to be a breakthrough. (Note: John Atanasoff's original computer at Iowa State College contained vacuum triode based memory circuits which essentially used the same principle). As silicon fabrication technology and chip yield continued to improve in the late 1960s, it became obvious that large numbers of these circuits, each with a single digital "bit" (the name for the "0" or "1" state) of information stored, could be integrated onto a single chip. Compared to magnetic core technology, silicon DRAM and other types of memory chips far outstripped their predecessors in terms of cost, size, power consumption, speed, and scalability. Soon, most major semiconductor companies began an urgent development effort to position themselves well in this new, potentially huge market.

In 1968, just one year after Dennard's breakthrough, Noyce and Moore left Fairchild to create Intel. They correctly assessed that there were no incumbent companies with any competitive advantage in the potentially huge computer chip memory market, and they determined that this was therefore the right time to start a new company specifically focused on a memory chip product. Within one year after Intel was founded, the company introduced the world's first 1,024-bit, or 1 K-bit, DRAM chip. Thanks to this technical feat, their little start-up became the world's de facto leader in the industry. Once Intel took the stage, Wang's magnetic core memory all but disappeared, completely annihilated by the advancement of the chip.

The computer industry's demand for high-speed memory chips proved inexhaustible. Because of greater memory capacity and faster speed, computers could be designed and developed to support ever more powerful and complex applications. For the general public, the most direct visible evidence of such advances came in the continuous improvement of the quality and resolution of computer graphics, from coarse outlines of shapes to real-life like objects.

Over the next several years, the memory chip industry grew rapidly, and Intel grew along with it. Noyce and Moore's company continually came out with new, groundbreaking memory chips, starting with the 1 K-bit, increasing to the 4 K-bit, and then reaching the 16 K-bit and so on. The number of transistors on each new chip increased exponentially, following Moore's Law. By 1975,

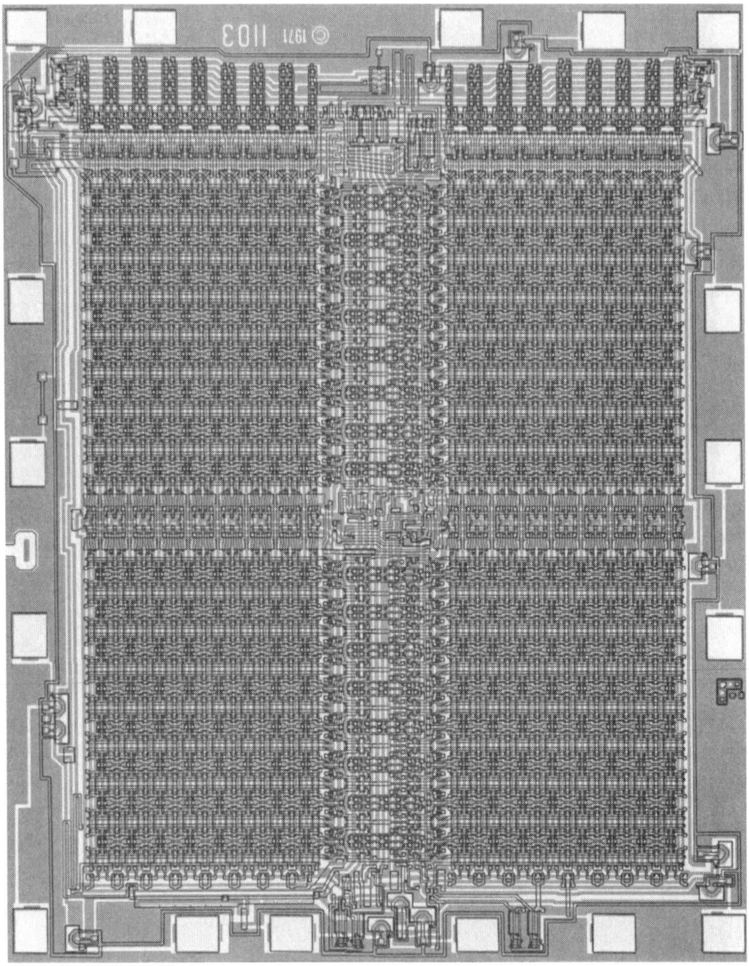

The first 1K memory chip. Courtesy of Intel Corp.

the top memory chips exceeded 10,000 transistors; ten years later, they were approaching one million. Today, DRAMs with capacities of multiple gigabytes (GB, or billions of bytes, each byte being a series of eight bits) are commonplace. DRAMs have become a worldwide commodity, and their day-to-day prices are quoted on trading boards together with other fungible goods like cotton, gold, pork bellies, and frozen concentrated orange juice.

In addition to DRAM, there exists a vast array of other specialized memory chips, including Flash memory, which is probably the most recognizable, as it is used in many of the consumer electronics that touch our daily lives, such as digital cameras, and USB thumb-drives. The digital information stored in a Flash memory chip is "nonvolatile." As opposed to DRAM, which needs to be refurbished continuously, the information stored in a Flash memory chip is practically permanent. The first patent for a Flash memory chip design was filed in 1980 by Fujio Masuoka, a Toshiba engineer who took it upon himself to develop the technology based on concepts first demonstrated by Bell Labs researchers Dawon Kahng and S. M. Sze back in 1967. Toshiba never took action to commercialize the Flash memory chip—or to give Masuoka credit—but Intel successfully introduced the world's first Flash memory chip in 1988 and reaped most of the benefit. By 2011, worldwide annual sales for Flash memory chips had reached $23 billion. More and more, our photos, videos, music, and other touchstones of our lives are recorded as clusters of electrons stored in these lattices of silicon atoms.

THE MICROPROCESSOR—ENIAC ON A CHIP

As important as semiconductor-based memory chips are to our modern life, they're only one piece of the puzzle, and they never would have reached the place of prominence they enjoy today had it not been for the invention of the microprocessor chip.

In 1954, armed with the new transistor technology, TI's Pat Haggerty masterminded the pocket radio, bringing TI into the enormous consumer electronics market. Though the radio itself was quite popular, market success eluded the company, and after a few years, TI retreated from the consumer electronics arena. It wasn't until 1968, when the new chip technology began to show promise for consumer applications, that the company made a second foray into the space, launching a small, handheld calculator powered by a handful of specially designed chips. For TI, this was a flagship project. As a sign of the company's commitment, TI appointed Jack Kilby, their best engineer, to be the technical lead. Thanks in no small part to Kilby's strong technical leadership, the chip-powered pocket calculator became an instant hit upon its introduction. One unique and useful feature that all users applauded was its integrated, miniaturized thermal printer, which was actually a Kilby invention. The new pocket calculator quickly became both a moneymaker for TI and a threat to the far bulkier and more expensive transistor-based desktop calculators. As a result, desktop calculator manufacturers began an urgent search for a counter to this new challenge. One such manufacturer was a medium-sized firm called Nippon Calculating Machine Company, better known by its product brand name, Busicom.

In 1969, only one year after Intel was founded, Busicom came knocking on its door, hoping that Intel could help the company's fight to compete with TI. The Japanese company had detailed a circuit design calling for twelve separate chips that they hoped would provide the core for a new generation of high-performance handheld calculators, ones that would be able to perform higher-order mathematical functions with ease. Busicom, supported behind the scenes by a rising corporate star in the Japanese electronics industry, Sharp, approached Intel to help them refine their design and fabricate the twelve chips, or chipset, using CMOS technology.

While it made sense for Busicom to approach Intel—the company was, after all, the chip-making industry leader—it wasn't clear that a partnership with Busicom was at all in Intel's best interests. In 1969, Intel was still a new start-up. Their primary focus had to remain on bringing out their own pioneering memory chip products. However, they did need the influx of cash this new business opportunity brought, so in the end, Intel reluctantly agreed to take on this work and signed the contract, assigning senior design engineer Ted Hoff to be in charge of the project.

Hoff was a seasoned engineer and an expert in computer architecture, with a PhD from Stanford. He studied the Busicom design, and after some careful thought, made a bold counter-proposal to his Japanese customer. He suggested that the original twelve-chip design could be reduced to only four chips, of which one—the "core"—would basically be a general purpose computer, or a central processing unit (CPU), combining high-speed memory functions with logic and arithmetic circuits. The various calculator functions, rather than being performed by fixed hardware operations, could be programmed in via computing instructions—in other words, software, just as von Neumann had envisioned years earlier on the ENIAC/EDVAC project. Hoff also assured Busicom that despite the relative youth of the company, Intel's chip manufacturing technology was already advanced enough that this "computer on a chip" product could be produced with reasonable yield. But the idea was so radical and risky that Busicom hemmed and hawed, unable to make up their minds whether or not to pull the trigger and proceed with Hoff's never-before-tested design.

As time passed with no definitive word from Busicom management, Intel's interest in this project waned. Eventually, though, seeing no alternative as TI continued to gain market share, Busicom hesitantly agreed to give Hoff's proposal a try. But by this time, Hoff's focus had shifted. He was busy with many other high-priority product development projects at the still-embryonic Intel, and he could only spend a limited amount of time on this contract work. This made progress painfully slow, and under such circumstances, the schedule for Busicom's product plan naturally slipped.

Busicom's market window was small and rapidly closing. The Japanese company's management was under tremendous stress, and despite their previous reticence, they kept pressuring Intel management to take more concrete actions to finish designing and producing the chipset. Eventually, Intel came to terms with the fact that in order to execute such a technically sophisticated venture with a foreign business partner, they needed a top-notch designer and a strong project leader who could manage the task independently without distracting from Intel's core business. This took Hoff off the table, since he was indispensable to other important Intel projects. The company needed a new champion—and who should arrive at just that moment but Federico Faggin, the outstanding Italian engineer from Fairchild Semiconductor.

Equally well known for his technical excellence and his combative temper, Faggin arrived at Intel already prepared to seek out his flagship project. Though it was not a part of the company's mainstream business line, the calculator venture with Busicom aroused his interest: the creative technical challenges excited him, and the relative independence of the project was right up his alley. What's more, since Faggin had not already integrated himself into other work flows, he was expendable. Intel management was therefore pleased to have Faggin to lead this project. This alleviated the pressure from their Japanese customer, and with Faggin working on his own, they wouldn't have to worry about potential conflicts between this newcomer and the rest of the staff.

After Faggin took charge of the project, he attacked it with a full head of steam. Hoff had left him conceptual notes and the overall architecture for the chipset—despite the amount of time that had elapsed, the team had not started detailing the chip design at all. This, however, didn't deter Faggin. He put together a small, crack squad, including a number of excellent Japanese engineers from Busicom, and they worked day and night with incredible intensity and dedication.

Faggin was a demanding leader. Every minute detail of his team's design was painstakingly worked to perfection. This level of focus slowed design work, but when the chip was at last completed in 1971, it was immediately clear that the team had done something spectacular. The finished product, internally named the 4004, was the world's first computer-on-a-chip, also known as a microprocessor. The size of the 4004 chip measured only 0.32 cm x 0.42 cm, and contained about 2,300 transistors. The computing power of the tiny 4004 chip was roughly equivalent to that of the ENIAC, even though the 4004 only consumed a paltry 1.2 watts in operation compared to the 160,000 watts needed to power the ENIAC—which also weighed 60,000 pounds! It took only twenty-five years to progress from ENIAC to the 4004. No one in 1946 could ever have dreamed of such dramatic technical progress in such a short period of time. It was the

The first microprocessor chip—the 4004. Courtesy of Intel Corp.

most dramatic manifestation of the power of the development of the transistor, and later chip technology.

Although the 4004 was a technical success and was indeed mass produced, it came too late for Busicom to make a splash in the calculator market. Not only had they missed their window, but their funds were exhausted by the lengthy development process. So despite Faggin's success, Busicom was forced to terminate the project. Since Busicom paid for the project, according to the contract agreement, they owned the rights to the intellectual property (IP) for the 4004—IP of which they had no idea of the true worth and that they were prepared to let gather dust on a shelf.

Faggin lobbied hard to turn the 4004 and its derivatives into standalone products; however, Intel's management had made a strong commitment to maintain the company's sharp focus on memory chips, so the 4004 didn't seem to fit with their business plan. At the same time they sensed that the 4004 might represent a significant yet undefined future opportunity. In a shrewd business move, Robert Noyce made a special trip to Japan to repurchase the ownership of the IP for the 4004 from Busicom for $60,000, with the caveat that Intel would be able to use this technology for all applications except to compete with Busicom in the calculator market. With this timely action, Intel kept open the option to build a new microprocessor-based business in the future with a very modest investment.

THE PERSONAL COMPUTER UNLEASHED

In 1974, armed with the support of Noyce and Hoff, Faggin and his team proceeded to design a more advanced and powerful microprocessor, the 8008. The 4004 was a 4-bit chip with very limited utility, only suitable for use in applications such as simple industrial controls, traffic lights, and programmable calculators. But the 8-bit 8008 chip was significantly more capable and sophisticated for a broader range of "embedded" digital signal processing applications, including performing some of the functions of a small-scale computer. This potential attracted a budding teenage software programmer by the name of Bill Gates, who attempted to develop a basic programming language to be executed on the 8008 chip; however, the effort was not successful because the chip was still not powerful and fast enough for the job.

Based on market feedback and their accumulated experiences, Faggin and his team soon developed their third microprocessor chip, the 8080, which turned out to be a great commercial success. In 1975, right after the chip's introduction, a small company called MITS built the first ever "microcomputer"—the Altair 8800, using the 8080 chip as its CPU. This time, Bill Gates and his partner Paul Allen were successful in developing their working BASIC (Beginner's All-purpose Symbolic Instruction Code) language for the 8080, and they leveraged this success to start a new company: Microsoft. Quickly, the prospect of a new microcomputer business looked real.

Following in the footsteps of Intel's pioneering effort, Motorola, TI, Fairchild, and others also entered the microprocessor chip market. Some of them leapfrogged Intel and introduced more advanced chips based on 16-bit architecture. In 1976, using Motorola's microprocessor chip, Steve Jobs and Steve Wozniak developed the Apple II computer, which ushered in the new personal computer age and threatened to unseat Intel from its market-leading position. Facing immense challenges from its competitors, Intel reinvented their business model. In addition to being a premium chip supplier, Intel invested heavily to build up a computer system architecture capability. Together with added software expertise, they offered what was known as "complete products" solutions to their customers. They pioneered the "product roadmap" concept, which allowed their customers to have a better view about the expected performance levels and time of availability for Intel's future microprocessor chips, which allowed these companies to have better product planning ability for the future. Of course this process also helped Intel to lock up their continued business.

With the incredible success of the Apple II computer, IBM, the giant of the computer industry, decided that they could not sit idly and let other new companies take over the growing market. In addition to throwing their weight around, they also needed to move quickly and decisively in order to define and dominate the new market. To follow this strategy, IBM set up a separate, nimble, and autonomous product team to capture the emerging microcomputer market, and they decided to focus the team's efforts on product definition and integration. To speed up the development time, IBM made the fateful decision to leverage the use of third-party microprocessor chips and software operating systems.

In 1981, IBM introduced its trend-setting entry to the microcomputer market and appropriately named their new product the personal computer, or PC. The IBM PC was a great business success, and it elevated the status of personal computer from specialized technical hobbyists' home use to legitimate office applications in the business world. Interestingly, in each PC, the two most important core technologies were not owned by IBM: the software operating system belonged to Microsoft, and the microprocessor chip came from Intel. Furthermore, IBM did not effectively restrict Microsoft and Intel from selling their products to other computer companies. As a consequence, the decision created a gigantic market for PC-compatible computers that was exploited by both new and existing companies, such as Dell, Compaq, and Hewlett-Packard. In hindsight, IBM's PC strategy may have been a big business mistake for the company, but it opened the floodgates to the PC revolution, which allowed the tiny electron to impact our lives in a way that is almost unthinkable. Along the way, Intel's foray into microprocessors and Microsoft's leap into software would forge the keys to an astronomically valuable treasure chest for both. Later, the Intel microprocessors also enabled the *servers*, such as those pioneered by Cisco, which was the building foundation of the incredibly powerful Internet revolution.

UBIQUITOUS SILICON

Building upon the twin foundations of the planar process and CMOS transistor technology, the range of functionality of silicon chips broadened rapidly. First there were logic chips, then memory chips, and then, Intel combined the two functions onto a single chip to create the microprocessor, elevating the industry to a new level and enabling the development of computers as we know them. Still, there are many other types of silicon chips that have both served and created a multitude of system applications, including analog chips, mixed-signal chips, imaging chips, power chips, micro-electromechanical systems (MEMS), and even super chips that aggregate all these functions onto one chip: Systems-On-a-Chip, or SOCs.

Analog Chips

All of humans' sensory interfaces with the natural world around them, including the perceptions of vision, hearing, and touch, are based on analog or continuously varying signals. The invention of the telephone for voice communication in 1876 ushered in the analog age in electronics. These analog signals are far more complex to transmit and process than the simple on-or-off dual states of digital signals.

Analog circuits were first used extensively in products such as the telephone, radio, and television, powered by vacuum triodes. Later, triodes were replaced by discrete transistors. But from the early days of chip development, there was always the desire to create an entire analog circuit on a single chip. Eventually, companies like Fairchild Semiconductor and National Semiconductor were able to introduce many successful analog chip products that embodied a range of key analog functions, from amplifiers and oscillators to frequency tuners and frequency selective filters. These chips upgraded the reliability and compactness of many existing analog products in communications and consumer electronics.

The number of transistors required on each analog chip is typically much lower than on equivalent digital chips. However, the performance specifications of each component are much more stringent and precise. Though the total market size for

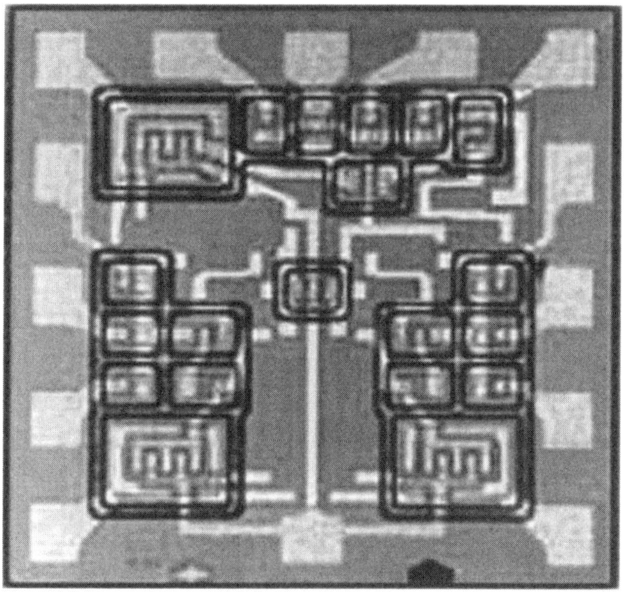

First analog chip: a linear amplifier. Courtesy of Fairchild Semiconductors

analog chips is sizeable, it is far more fragmented than the digital chip market and is served primarily by highly specialized chip companies.

Mixed-Signal Chips—Data Converters

In our lives, all signals that we directly deal with are analog in nature; however, in transmission or storage, analog signals are vulnerable to distortion, attenuation, and noise contamination. In contrast, digital signals are much more robust, and they can be readily regenerated, as was first demonstrated by the early telegraph relays. There is another significant advantage to digital signals: using computer functions, they can readily be "digitally processed" with a wide variety of powerful software-based mathematical algorithms, enhancing their value to the user. For example, a digitized picture can be processed to enhance its contrast, color, and sharpness; the picture can be "compressed" for storage or for ease of transmission; or for more advanced processing, the computer may even have the intelligence to recognize special features in a picture and take actions, such as Facebook tagging of individuals in photos thanks to facial recognition.

To take advantage of the properties of digital signals, analog signals must be converted into a digital equivalent, which is represented by strings of 0s and 1s. The more "bits" used to represent the original analog signal, the higher the accuracy of the digital approximation. To a computer, Beethoven's symphony and Da Vinci's paintings are nothing but combinations of 1s and 0s—and the quality of the image redisplayed or the music replayed will inherently depend on the bit-depth of the digital encoding process. Thus, improving methods of converting signals back and forth between analog and digital states have been significant technical goals ever since the early days of the computer.

The emergence of the microprocessor created an ever greater urgent need for low cost, high-performance data converters which competed to perform these conversion processes at the highest possible speed and with the highest accuracy. To address the need, a new chip technology was developed that integrated both digital and analog functions onto the same chip. This combined capability became known as "mixed signal" technology. As chip functions continue to aggregate, data converter functions become ever more integrated with sensors, actuators, and microprocessors, making the gap between the analog and digital worlds effectively transparent to end users. Thanks to these mixed signal data converters, the power and value of microprocessors has truly been unleashed, creating a new generation of digital products ranging from CD players to intelligent rice cookers.

Mixed Signal Chips—Wireless

One of the benefits of Moore's Law was that as transistors' size shrank, their maximum operating speeds accelerated, eventually becoming fast enough to effectively amplify wireless signals used in mobile phones with frequencies of two gigahertz (two billion cycles per second) and beyond. The operating speed of microprocessors, too, followed in lockstep. This increase in speed led to the development of wireless transmitters and receivers integrated onto a chip, a technology that enabled compact, low-power, portable wireless products like the cell phone, the GPS receiver, and Wi-Fi and Bluetooth devices. The low cost of CMOS chip manufacturing allowed these products to become easily affordable to the general public. Their widespread adoption ultimately ushered in a new generation of smart phones and mobile applications.

Imaging Chips

In 1969, Willard Boyle and George Smith of Bell Labs were studying the interface between silicon and silicon dioxide when they made an interesting observation: applying positive voltage to a gate electrode over p-type silicon would cause the holes to be repelled away from the silicon/silicon dioxide interface. That space under the gate electrode could be used as a storage "bucket" for the free electrons generated there by light absorption.

Using a linear array of closely spaced gate electrodes, Boyle and Smith elegantly demonstrated that electrical voltage, applied in sequence to the gates,

The first 8x1 CCD chip. Reprinted with permission of Alcatel-Lucent USA Inc.

could physically move packets of stored electrons around these buckets in a controlled fashion. The device they created to take advantage of this phenomenon was named the charge-coupled device, or the CCD. Very soon, they also developed two-dimensional arrays of CCDs.

Boyle and Smith's original idea was to use the CCD as a new type of electronic memory, using the stored electron packets under each gate to represent 0s and 1s. However, though the CCD could perform this function, further experimentation and analysis proved that the CCD could not compete with DRAM in terms of storage density or power consumption. They had failed to achieve their goal of creating a challenger to DRAM. However, in time, they found that their invention's specific attributes made it perfectly suited for applications in electronic imaging.

When an image is projected onto a two-dimensional CCD array, the silicon absorbs the light, generating electrons and holes. Holes are swept into the p-type silicon material and conducted away, but electrons are attracted to the silicon/silicon dioxide interface and accumulate in the "buckets" under each electrode. The amount of electrons stored under each gate is proportional to the incident light intensity and exposure time, and each electrode defines a point, or pixel, on the two-dimensional matrix of the CCD chip. Thus, the spatial distribution of electrons accumulated under each gate of the CCD is an exact replica of the optical image captured, and this two-dimensional electron distribution pattern can be immediately converted into a digital serial current for output or storage in memory. These sequences of currents can be copied, reproduced, and used to reconstruct the original image ad infinitum using an electronic display or a printer.

The CCD provides outstanding imaging capability and extremely high sensitivity. The signal readout rate is also fast, allowing easy adaptation to video applications. What's more, the light absorption characteristics of silicon match well with those of the human eye, so by using color filter arrays over each pixel and with simple recalibration, even the colors of the captured images can be readily reproduced or enhanced.

Though the CCD failed to function as a viable memory device as Boyle and Smith had originally hoped, it ended up proving capable of so much more in the field of electronic imaging. In fact, in 2009, Willard Boyle and George Smith were awarded the Nobel Prize in Physics for their work on the CCD, their revolutionary failure!

The basic production technology for the CCD was very similar to that used in CMOS chips for memory and microprocessors, so it was quite easy to volume-produce CCDs in already-existing chip production facilities. Sony, in fact, became a leader in CCD technology for many years, and they leveraged that advantage to become the dominant supplier of high-end electronic television cameras.

Over the decades since their invention, the maximum resolution (measured in number of pixels) of CCDs increased in lockstep with Moore's Law, just as microprocessors and memory chips have. Today, ultra high-sensitivity CCD imagers boasting over ten million pixels are commonplace, even in mobile phones. Surveillance electronic camera systems are now so ubiquitous that the society where Big Brother is watching over us all the time is already here!

In recent years, a new CMOS-based imaging technology has been developed to challenge CCDs, especially in cost-sensitive applications such as low-end cell phone cameras. Together, CCDs and the new CMOS imaging technology have revolutionized digital photography, making chemical-based photographic films largely a thing of the past.

Power Chips

Most of the applications for silicon chips deal with communication and information processing, and these chips typically operate at very low power levels. But there exists another class of semiconductor devices designed specifically to handle and manage extremely high levels of electrical load. These power transistors and diodes are used in high power electrical systems, such as power grids and electrical trains. One particularly important silicon device for power applications is known as the IGBT, or Insulated Gate Bipolar Transistor. IGBTs are used extensively in high power switching operations, which enable a broad range of variable-frequency motor drives that form the core of a new generation of highly optimized, energy efficient applications, such as EnergyStar-compliant air conditioners.

Micro-Electromechanical-Systems (MEMS)

Silicon chip fabrication requires an extremely high degree of precision and control. In the early 1990s, researchers leveraged high-precision photolithography, etching, and thin-film technology to fabricate miniaturized electromechanical systems on silicon chips with integrated moving parts. The development of MEMS added a new dimension of functionality to silicon chips. Many impressive designs have been created, including high-speed miniaturized motors, accelerometers, and gyroscopes. These products are now ubiquitous—for instance, they are embedded within every smart phone, allowing the devices to sense motion and orientation. Commercially available MEMS-based products also include chip-scale image projectors with two-dimensional arrays of

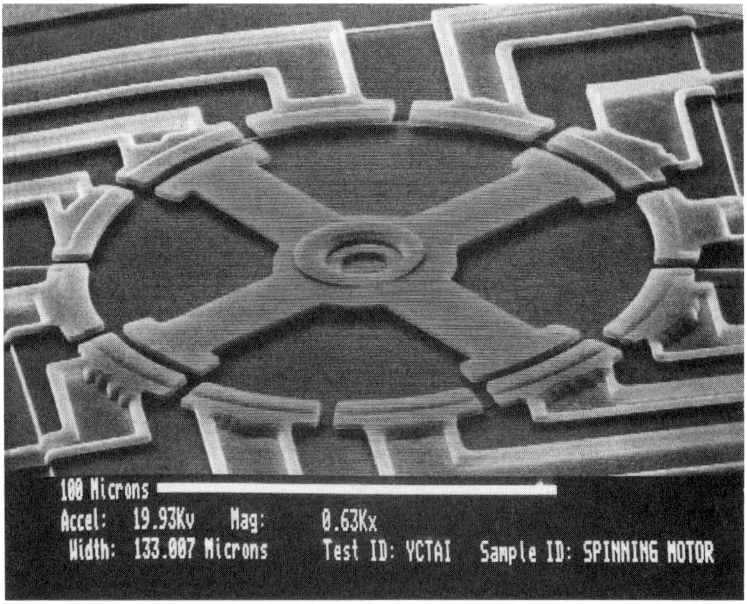

Micro-motor integrated on a MEMS chip. Courtesy of Professor Y. C. Tai

steerable micro-mirrors, miniaturized microphone arrays, and a broad range of chemical and biological sensors. They hold the key for future generations of fully integrated, human-implantable biosensors.

Systems-On-a-Chip (SOCs)

Throughout chip development history, one axiom has consistently proven its worth: it is advantageous to integrate as many circuit functions onto a single chip as possible. As Moore's Law evolved, manufacturing technology continued to improve, enabling ever more complex chip designs to take shape. In time, it even became possible to integrate an entire system's capability on a single chip, including both the hardware and stored operating software codes. This new class of super chips became known as the System-On-a-Chip, or SOC. Generally, these highly-integrated SOCs succeeded in reducing overall end product cost while enhancing performance and reliability.

The first SOC was the 4004 microprocessor chip conceptualized by Ted Hoff and built by Federico Faggin. After Intel's microprocessor-on-a-chip, TI developed a calculator-on-a-chip, followed by a watch-on-a-chip. In the 1980s, Rockwell International ushered in the era of modern data communications by developing the modem-on-a-chip, which transmitted and received digital data through analog telephone lines and was widely used in fax machines.

In subsequent years, this trend of situating entire systems on a chip has only continued to accelerate. Combining wireless chip technology with the microprocessor, we now have the cell-phone-on-a-chip. By integrating an imaging chip with a microprocessor, we now have the camera-on-a-chip. Consolidating a MEMS-based accelerometer and gyroscope with a GPS receiver and a microprocessor, even full-fledged navigation systems are now available on a single chip. The trend in recent years is to further integrate these SOCs into a "system of systems," which form the core building blocks of a new generation of compact, lightweight, low power consumption smartphones and tablets. As long as there remains an advantage to aggregating the functions, provided the design and fabrication technologies can support operational requirements, there will be no end to this march. Along the way, the products that are enabled by these SOCs have become more multifunctional, more compact, and more affordable.

The design of SOCs is extremely complex and requires significant system knowledge and expertise. In most cases, SOC suppliers control the core system technology and the intellectual property. End product manufacturers effectively become construction workers, assembling the SOCs and other peripherals as building blocks toward completing the final product. This trend has created an inverted supply chain in which SOC suppliers own most of the cutting-edge technologies and wield great influence and pricing power over their downstream product manufacturers and distributors.

Technically speaking, the development of the SOCs and the new system-of-systems chips have brought us to the current state of mankind's long journey to harness the power of the electron. These building block chips enable the smartphones and tablet computing platforms that can perform many useful and essential tasks, enriching our daily lives. Only ten to twenty years ago, most of these capabilities were not even mere dreams; they had never even been imagined to be possible.

18

EVOLUTION OF THE ELECTRONICS INDUSTRY

COMPETITORS FROM ASIA

Through the 1970s and early 1980s, the worldwide semiconductor industry was full of vitality and growing rapidly. Like mushrooms after a storm, innumerable start-ups popped up on the scene, each one introducing new chip products and addressing novel markets.

Following the lead of Sony, starting from the 1970s, Japanese companies began to make inroads into the global semiconductor industry. These big Japanese enterprises were capital-rich, well-coordinated, and supported by a pro-industrial national economic policy. Within a relatively short period of time, they birthed multiple world-class semiconductor manufacturing plants. Japanese engineers and production workers were well-trained, meticulous, and highly disciplined, all qualities highly favorable to the mass production of complex and high precision products. These companies also adopted Total Quality Management practices, making their competitive edge even more formidable.

The main product upstart Japanese companies took an aim at was the DRAM chips. The upsides were obvious: the market for this product was huge, yet the product design was relatively straightforward. New product cycles were predictable and lasted a reasonably long time. Furthermore, provided quality could be assured, there was relatively little differentiation between products besides price, and in that respect, Japanese companies had a leg up on the competition. The yen was weak relative to the dollar, making the market price of Japanese DRAM chips artificially low, and therefore attractive to buyers. By the early 1980s, Japanese DRAM products were threatening to corner the world market, and they began to pose a serious threat to other entrenched producers—including Intel.

The memory chip had been Intel's core product since the day the company was founded. But Intel now found itself in a fight against a group of powerful and determined rivals. Its market share began to drop, and profit margins squeezed. The company needed to act. Fortunately, thanks to Noyce's strategic vision and Faggin's obstinacy, after several years of development, Intel's microprocessor business had begun to take off in lockstep with the phenomenal growth of the personal computer market. Intel's microprocessor was the first and the best, and it was designed into most personal computers, including the industry standard: the IBM PC. As growth prospects for the microprocessor were extremely bright—far better than those for memory chips—in 1986, Intel made a strategic decision: the company announced that it was exiting the DRAM market and would concentrate solely on developing its microprocessor business. The U.S. electronics industry was shocked by this news. To outsiders, it appeared that, in ceding DRAMs, Intel had announced that Japanese players had crushed their U.S. counterparts in the electronics industry. However, this was not the case.

Because of its high growth rate, the semiconductor market was not a zero-sum game. In an ever-expanding market, there were opportunities for winners to exist across multiple points in the supply chain and with many types of final products; one company's success did not mean another company's failure. However, because there were so many players, each company had to be agile and flexible in this competitive and highly differentiated environment, and each firm also had to have a clear view of its own unique competitive advantages.

Intel was not immune to this reality. Their managers were able to cast off the blinders of the past and come to terms with the fact that, despite their historic dominance, they would not remain the global leader of DRAMs for long. The simple truths of international commerce would not allow it. However, they also perceived that they could lead the field in the microprocessor market: they had more experience and know-how than any other company in the field. Besides, the markets for personal computers and high-end "smart" products with embedded microprocessors were rapidly growing. They could attack these markets and stand secure, especially given the number of barriers to entry that faced potential would-be competitors: the complexities of design; the technical requirements of rapidly evolving hardware, software, and system applications; and the nuanced interplay required to forge multifaceted, cooperative relationships with customers. Intel's advantage was not their past strength; it was their newly acquired ability to skillfully adapt to market changes and transform themselves to address a new and growing market. In addition to being a semiconductor product company, Intel became a

A packaged micro-controller chip for smart digital products. Courtesy of Intel Corp.

leading systems company with an enormous wealth of expertise in computer architecture. As a result of their dominance in this field, they not only controlled most of the market, they also possessed pricing power. So for Intel to move in 1986 from DRAMs to microprocessors wasn't truly ceding ground; it was finding a newer, more profitable, and stronger footing for their long-term businesses.

In the years after Intel's decision, the company was able to quickly recommence high-speed growth, and they regained the title of the world's biggest semiconductor company. As for their competitors in memory chips, many found themselves fighting one another for what quickly became slim profits in a commodity market. When the Japanese economic bubble burst and investment in the semiconductor industry dwindled, new companies from other parts of Asia, including Taiwan, Singapore, South Korea, and much later, China, emerged, beating the Japanese firms at their own game. These days, for DRAM production, South Korea's Samsung Corporation has become by far the largest supplier in the world, and some of the companies that market-watchers feared Intel were kowtowing to at the time have long since exited the chip business altogether.

COMPUTER-AIDED DESIGN

As the semiconductor industry continued to grow, a bloom of innovations pushed the boundaries of production capability forward. Some were technological advances, while others were instead related to business methodology or division of labor in chip design and fabrication. These developments, too, had a significant impact on the evolution of the global industry.

In accordance with Moore's Law, cutting-edge chips grew to contain over ten thousand transistors by the mid-1970. Circuits with many differing functions had begun to be combined and integrated—electronic watches that had originally required five different chips, for example, could now be combined into one integrated system-on-a-chip. With all this added complexity, a new critical bottleneck emerged: chip design methodology.

Large scale integration (LSI) chip design was a highly specialized field, and most chip design engineers were initially employed at big semiconductor companies like Intel and TI. Qualified, experienced chip designers were few and far between: they not only had to know circuit design, but they also had to understand the limits of chip fabrication techniques and semiconductor physics. This predicated mastery of multiple fields and years of experience. In truth, the best of the best were more like artists than engineers. Top-notch chip designers such as Federico Faggin were extremely hard to find and even harder to hire—they were such precious resources that they tended to be employed only in the prestigious and high-pressure roles of designing their companies' most important products, and they were well-compensated. However, even among the few talents who possessed the requisite expertise, there was no shared book of knowledge. Everyone learned on the job, and each designer crafted his own set of experiences from which to draw; there was little standardization. What the industry needed was some way to develop a systematic methodology for chip design that would minimize the dependence on a limited number of chip designers and open up the field to a large population of engineers with modest training. Among the many researchers working to break through this design bottleneck, one major contributor was a man named Carver Mead.

Carver Mead was a professor of electrical engineering at Cal Tech. He had originally specialized in researching the physics of the interface between semiconductors and metals, a field first explored in the 1930s by Walter Schottky. But Mead was a man of broad interests and versatility, and when, during the mid-1970s, he saw the immense need to revolutionize chip design methodology, he realized that computers could play a central role.

Chip design methodology was a field wide-open for innovation and with enormous potential payoff. Even though this field of research was out of Mead's principal area of expertise, he felt no hesitation in pursuing it. He wanted to expand his horizons and make a major impact, so he attacked the matter with fervor.

Mead was an intellectual maverick, the kind of man who always approached problems differently than the mainstream. Rather than being hidebound by

Carver Mead. Courtesy of Carver Mead

the notion that proper chip design could come only through the artistry of years of multidisciplinary experience, he perceived that over time, some best practices would emerge and some design tactics would converge. As component technology in chip manufacturing matured, a growing number of component functions would become stable and reproducible. Therefore, since their utility was essentially known, they ought to be able to be modeled by computer simulation, just like any other mathematical representation.

Riding this method of thinking to its logical conclusion, Mead developed a methodology to convert any given logic system design into a succession of standard building block circuits that could then be further translated into a chip layout. Using software, he simulated the performance of those prospective chips to see if they met sets of necessary functional requirements and performance targets, at which point, if they did, they could be cleared for manufacture. If their simulated performance failed to meet specifications, he could simply refine and iterate the design until the desired results were achieved.

Computer-aided chip design was a revolutionary concept. In 1980, Mead and his partner, Lynn Conway, published a textbook named *An Introduction to Very Large Scale Integration Systems* that systemized the principles of digital integrated circuit design. The book championed the notion that, using Mead's methodology and appropriate software design tools, engineers, even without prior semiconductor knowledge, could become competent digital chip designers. Systems and end product–oriented chip users, seeing their potential, rapidly adopted Mead's ideas and began to set up their own internal chip design departments to develop special chips for their own needs. The bottlenecks were reduced, the value chain was righted, and the industry was free to grow smoothly and enjoy a rapidly expanding product variety.

As computer-assisted chip design gained traction, a new software industry was created. Many powerful chip design and simulation software tools became available, addressing a broad range of design needs. These tools played a critical role in enabling the layout of complex chips with millions of components without any errors or inconsistencies. The use of these tools also lowered the threshold to chip designers of developing myriad new chips with countless new functionality.

Although computer-aided design significantly reduced the cost and time required to design digital chips, building a brand-new, application-specific chip from the ground up was still a time-consuming and expensive investment. In response to this need, a new type of chip product was created: gate arrays. Conceptually, gate arrays were simply chips that contained modular, prefabricated logic circuit blocks that had not yet been connected to one another, much like children's Lego toys. By using special-purpose design tools, users could customize these intermediary chips for their own specific applications. They only needed to properly connect the assorted components and building blocks on the unfinished chip by using an interconnect mask generated by the computer-aided design tool. These interconnected patterns transformed function-nonspecific gate arrays into specialized chips with unique functionality. Soon, simple gate arrays evolved into field-programmable gate arrays or FPGAs for short. Rather than requiring the physical connection of its components, these FPGAs were already fully finished and connected. However, the routing or wiring of the components could be reconfigured by the user via software inputs. This capability expanded the breadth of gate arrays' applications even further.

Today, every chip is blueprinted with sophisticated design, layout, and simulation software before it enters physical fabrication. The complexity of these chips has long exceeded humans' capacity for systems design without the aid of a computer. To think that as recently as 1971, at the birth of the microprocessor, Federico Faggin and his team used their bare hands to design and layout nearly every component of the 4004 chip without making an error . . . it's all a bit unimaginable now.

THE FOUNDRIES OF TAIWAN

Once the chip design bottleneck was broken, the structure of the semiconductor industry started to change. Many formerly dependent companies—those at the end of the product supply chain—found themselves able to design special chips for their own use. They were no longer at the mercy of chip suppliers. That said, even if they could design their own chips in-house, they still needed semiconductor manufacturers to physically fabricate the chips for them.

At the time, almost all chip-centric companies had their own fabrication lines. For many of them, seasonal fluctuations in production demand could result in occasional over-capacity at their factories. Thus, in order to increase revenue and offset cost, some companies began to offer limited contract fabrication services to companies that had just started designing their own chips. The world of chip design and fabrication started to show signs of creating a division of labor.

In 1973, right after Faggin brought the microprocessor into the world, but before the computer-aided chip design revolution, Taiwan's traditional industries fell into hard times. The country's chief executive, Premier Chiang Ching-Kuo, decided that in order to keep the economy afloat, it would be imperative to develop new avenues of business. A group of advisors with foresight boldly proposed to him to develop a semiconductor industry in Taiwan, and based on that proposal, the government agreed to invest $14 million to lay the foundations of a semiconductor industry.

The plan was to assign to a quasi-governmental organization called the Industrial Technology Research Institute, or ITRI, the responsibility for selecting and licensing a semiconductor technology upon which to base Taiwan's electronic industry. ITRI would then receive the technology in Taiwan and have the responsibility to recruit talent, train a local work force, and further develop the technology, with the ultimate goal of transferring that technological expertise to private sector enterprise in Taiwan.

Pan Wen-Yuan, fresh from a successful career in RCA's semiconductor division, was coaxed out of retirement to helm the ITRI project. After careful analysis, he and his team decided to license the CMOS technology from Pan's old company, RCA. This was the right decision at the time, as CMOS had quickly emerged as the mainstream technology. However, it took ITRI the better part of a decade to successfully build up Taiwan's technical foundation and the skills of ITRI's core engineering team, meaning it wasn't until the early 1980s that the quasi-governmental agency was finally able to spin off several semiconductor companies. CMOS technology had advanced rapidly in the United States and Japan in the interim, and the fundamental CMOS technology Taiwan possessed had now aged and fallen two full generations behind. Thus, these early Taiwanese semiconductor firms enjoyed only very limited market success, and there appeared to be no clear path to build an industry that could offer any unique competitive advantage over other global competitors. This left the Taiwanese semiconductor industry in a quandary: what course of action should they take going forward?

In 1983, ITRI recruited a new leader: Morris Chang. Chang had been born into a middle-class family in China in 1931, and as a small boy, he and his whole family were uprooted by the devastation of the Sino-Japanese war and the Chinese Civil War. He saw turmoil, poverty, and injustice early on in life, and these experiences left a deep impression upon him. In 1949, Chang and his family moved to America. Being an excellent student, Chang was accepted to Harvard, but he transferred a year later to MIT, where he majored in mechanical engineering with special emphasis in automation technology. After receiving his BS and MS degrees, he planned to continue on to earn a PhD, but surprisingly, he failed his qualifying exams. Realizing that it might be time for a change, Chang decided to first find a job in the industry, and he was hired by Sylvania, one of the earliest licensees of Bell Labs' transistor technology.

Morris Chang. Global Views Monthly/Chen Chih-Chun

At Sylvania, Chang was assigned to work on transistor product development, where he was first exposed to semiconductor technology. He taught himself semiconductor physics by studying Shockley's masterpiece: *Electrons and Holes in Semiconductors*. But Sylvania was a very conservative company, totally lacking in vision and aggressiveness. Also it owed its original business success to vacuum tubes and the management never had the passion for transistor. Being an ambitious young man, Chang was frustrated by the sluggishness of Sylvania, and after two years, he decided to look for a new job.

In 1958, the same year that Jack Kilby signed on (for very similar reasons), Morris Chang became an employee of TI. Like Kilby, Chang performed brilliantly in this new, stimulating environment, quickly rising through the ranks until he became TI's senior vice president in charge of all semiconductor business. He also found time along the way to at last earn his PhD from Stanford University. But after a twenty-five-year career at TI, Chang reached the decision that it was time to move on. It was just at that time that an old friend called from Taiwan, and after more than three decades of living in America, Chang returned to Asia with the determination to positively impact the economy of Taiwan. Chang took over the reins of ITRI in 1983. He helmed the organization as best he could, but the technology they owned was indeed well behind the curve. At the beginning of 1987, a very important Taiwanese government official named Kwoh-Ting Li called on Chang with a strident demand for innovative thinking. Li was heralded as "the Father of Taiwan's Economic Miracle" for having had a hand in more than quadrupling Taiwan's GDP in the time since he had first been named to the top government post for economic development. Chang knew that meeting with a man of such stature meant that he was about to be tested as he never had been before.

As expected, Li made clear to Chang that he was concerned by the uncertain future of Taiwan's chip industry, especially given its lack of unique competitive advantage. He encouraged Chang to set up a new chip company, one that would somehow catapult Taiwan's semiconductor industry and strengthen its global position. As if this were not a tall enough order, he granted Chang only four days to develop a business plan! But Chang had left TI looking for a challenge, after all, and he did not shy away from the task.

Relying on deep understanding about the global electronics industry formed over a lifetime in the field, Chang flexed his Shockleyan "will to think," considering the unique strengths and weaknesses—and there were many of the latter—of the Taiwanese semiconductor industry. Eventually, he hit upon an idea.

"What the new company should do," Chang boldly proposed to Li, "is to concentrate its business on being a 'silicon foundry.'" In other words, Chang conceived of a company that solely provided chip manufacturing services to customers. It would be a service-provider, a manufacturer-for-hire, and it would not produce any commercial end products of its own.

The idea was nothing short of revolutionary at the time. It shifted the semiconductor industry from a product-centric culture into a service industry. By cutting out the step of developing branded products in-house for general sale in the marketplace, Chang removed the potential for the company to run into conflicts of interest with its customers. Functioning exclusively as a manufacturer, the company would focus solely on serving its customers' needs, and it could ensure the ability to safeguard the proprietary designs of each customer. The foundry's ultimate main customer base became the "fabless" companies—the contingent of firms that designed and sold their own chips but did not physically manufacture them in-house. In 1987, this niche was small since the computer-aided design methodology was not yet in widespread use, but Chang anticipated that, as the industry grew, more and more companies would employ this rubric as they attempted to differentiate. In such a market, division of labor would be a key to success. After all, the companies who first succeeded in the transistor industry were not the megalithic corporations like GE, but smaller upstarts focused on doing one thing only and doing it right. That was what the new firm should focus on: chip manufacturing. It was a big bet, but Chang's pitch was convincing. Li acquiesced, fostering the birth of the Taiwan Semiconductor Manufacturing Company, or TSMC.

Chang's instincts and timing were impeccable. Indeed, more and more fabless companies entered the chip marketplace as the computer-aided design technology matured, and, no longer beholden to the scheduling inflexibility of large chip-makers or the danger of compromising their own proprietary chip designs, they turned to TSMC. Under Chang's leadership, TSMC went on to become the largest and best silicon foundry in the world, with 2013 revenue over $20 billion. It was known for offering the best customer service and providing the most advanced chip fabrication technology available. Eventually, other pure foundry companies followed TSMC's lead, serving as a rising tide that lifted all boats: they helped inspire a proliferation of fabless chip companies, which in turn drew in a swell of customers. This once again changed the supply chain of the entire electronics industry, but no other company ever really challenged TSMC. Through it all, TSMC stayed on top.

As time progressed, transistors became smaller and smaller, while the maximum size of wafers—the thin discs of rounded, single-crystal silicon material atop of which chips were built—expanded. As a result, each wafer could accommodate more chips, which again led to lower chip cost. The diameter of early silicon wafers was only 1 to 2 inches in the 1950s, but reached 4 inches, then 6, then 8, then 12 inches. In 2014, the diameter of the largest silicon wafer reached 18 inches. That means that factories will be churning out perfect, brittle discs in sizes larger than an extra-large pizza, and each wafer will contain thousands of chips! Of course, when a new generation of wafer size is adopted, production facilities and equipment must be upgraded, and setting up a new state-of-the-art silicon chip factory requires an investment of multiple billions of dollars. The operational costs of such facilities are high as well, as is the rate of depreciation of equipment. Only a very few well-funded electronics enterprises, such as Intel, TSMC and other first-tier silicon foundries, can economically justify such capital investment. The business model for foundry companies is to stay at the cutting edge of chip fabrication technology and share their capabilities, capacity, and associated costs with many customers. This highly effective business model is a keystone to allowing spectacular growth of the modern chip industry.

NOYCE, MOORE, AND GROVE

History showed that Intel's business decision to leave the DRAM market was a correct one. Thanks to the company's entry into microprocessors, Intel was able to highly influence the direction of the personal computer business, much as Edison-GE had shaped the evolution of electrification. And just as the electrification process and the subsequent build-up of electrical power and transportation systems were largely influenced by key pioneers such as Edison, Tesla, and Siemens, the equivalent key figures of the silicon chip industry were the heroic trio of Robert Noyce, Gordon Moore, and Andy Grove.

To review: in 1955, the first two of these young engineers were recruited to and joined forces at Shockley Semiconductor Laboratory. In 1957, they founded Fairchild Semiconductor as core members of the Traitorous Eight, planting the seed for the growth of Silicon Valley. Eleven years later, in 1968, they left Fairchild Semiconductor with Andy Grove to found Intel, a start-up that would grow to become the largest and most successful semiconductor company in the world.

Robert Noyce was born in Burlington, Iowa, in 1927, the son of a minister. As a child, Noyce was intelligent, athletic, and charismatic. He had wide-ranging interests and was gregarious, and his leadership abilities were abundantly clear even from an early age. When he joined his school's Glee Club, he became both its leader and conductor; when he joined the swimming team, he was elected as its captain. Everyone liked to be with him and looked to him for guidance.

Robert Noyce. Courtesy of Intel Corp.

As it happened, Burlington played host to an excellent liberal arts college, Grinnell, and after graduating from high school,

Noyce enrolled there and majored in physics and mathematics. In 1949, Noyce's physics professor, Grant Gale, obtained several point-contact transistors from a personal friend, a very important scientist at Bell Labs: John Bardeen. As Professor Gale demonstrated and lectured on the physics of the new transistor, the young student became totally mesmerized by the subject. Upon graduation, Noyce applied to and was accepted at MIT. He moved to Boston to study there, and he earned a PhD in semiconductor physics in 1953.

Despite his demonstrated excellence in academics, Noyce was a highly pragmatic person, and his objective was always to work in industry. His first job was with the Philco Company, an early licensee of Bell Labs' transistor technology. There, Noyce was able to make some improvements to transistor technology, and published several papers in technical journals that caught the eye of Shockley. For all his talent and success, though, Noyce was certain that his future at Philco was limited—it was an old-fashioned and conservative company, one lacking any concrete plan of building up a new business in the nascent field that so interested the young man. (This was the same career dilemma that faced Jack Kilby and Morris Chang.) Thus, when Noyce received a call from Shockley in 1955 inviting him to California, he packed his bags without hesitation. Noyce even bought a house in Palo Alto before receiving a formal letter of employment! But then, Noyce was always impetuous. At Grinnell, he'd stolen and slaughtered a pig for a campus luau without considering the possible ramifications. He was caught, tried by a campus tribunal, and in fact, faced expulsion. But Professor Gale argued vehemently that Noyce's punishment be commuted so as not to forever waste the young man's undeniable brilliance.

At Shockley Semiconductor, Noyce joined a group of outstanding young men who were in many respects his equal. Indeed, every employee had just one thing in mind: working with the world's foremost expert to build the best transistor products. Still, even Shockley was impressed by Noyce's uncommon competence, so he always treated Noyce a little differently. Because of this, Noyce was initially reluctant to join the other seven rebels, out of his respect and loyalty to Shockley. Of course, he eventually did, and his indomitable spirit would be recognized when he was tapped to lead his team at the next company he joined as well.

In 1957, Sherman Fairchild invested $1.5 million to start Fairchild Semiconductor. Afterwards, he told his associates that what had impressed him most was Noyce, especially the business insight and the enthusiasm that Noyce brought to his investment presentation. After Fairchild Semiconductor opened for business, Noyce gradually became the leader of the firm, and once Fairchild bought out the Traitorous Eight, the corporate executives charged Noyce with reporting directly to the parent company.

Noyce's creativity and innovativeness came from an inexhaustible wellspring, but his energetic impetuousness, one of his greatest strengths, was also his Achilles heel: it often kept him from being able to carefully sift bad ideas from good ones or to always see good ideas through to their conclusions. However, Noyce recognized his weaknesses. He knew that he couldn't always bring perfect scrutiny to bear, and because of that, some of his ideas would emerge from his mind incomplete or even dead wrong. Noyce didn't let this stop him, though, because he had a solution to his impetuousness problem: Gordon Moore.

Gordon Moore was Noyce's most trusted partner. Whenever Noyce wasn't sure about some conclusion he'd reached, his preferred remedy was to discuss the matter with Moore. If his concepts could hold up under Moore's careful scrutiny, Noyce knew that he could proceed without hesitation.

Gordon Moore was born in San Francisco in 1929, and he grew up in Pescadero, a Northern California suburb where his father was the lone local sheriff. Moore started college at San Jose State University, but he soon moved on to UC-Berkeley, where he majored in chemistry. He later got his PhD in chemistry from Cal Tech.

Gordon Moore. Courtesy of Intel Corp.

Moore's first job out of school was at Johns Hopkins University's Applied Physics Laboratory in Maryland, where he did research on military technology. Though the position was prestigious, Moore found that he was uneasy with the job: he believed that many of the projects had questionable value, that the productivity of the lab itself was low, and that many of the staff members lacked motivation. He didn't want to waste his abilities and time just for a salary; he wanted to do something impactful. After just a year at the Applied Physics Lab, Moore began to look for other opportunities. He was offered a job at the Lawrence Livermore National Laboratory back near his hometown in Northern California, but he turned it down—the position was just too similar to the one he already had.

Just as Moore was starting to wonder what would be the next step for his career, he was contacted by Shockley, who found Moore's resume through Lawrence Livermore National Lab. Fate had its way: Moore joined Shockley Semiconductor, then left to start Fairchild, then left that company, too, to found Intel with Robert Noyce.

Though Moore was extremely intelligent and meticulous, he was also patient and modest. His favorite hobby was fishing, one of the most passive sports. It is said that in most meetings, he verbalized very little, preferring to listen attentively to everyone else's comments. At the end, he would utter a few words, which were nearly always concise, on target, and actionable.

In 1968, Noyce and Moore left Fairchild Semiconductor and founded Intel. At the helm of their own company once more, they were determined to make it a success and to avoid repeating the past mistakes of Fairchild.

From the beginning, Noyce and Moore established a disciplined management style at Intel. They also implemented some innovative business practices, including integrating research activities with the production operation to avoid the pitfalls commonly associated with technology transition. At the same time, they partitioned knowledge of core technology between groups to prevent intellectual property loss, which had been a common problem at Fairchild with its constant exodus of talent. Their deep understanding, insight, passion, and vision into the technological drivers of their field made them all the more valuable when compared to traditionally trained, finance-centric corporate managers.

While Noyce and Moore established strong corporate discipline, they also avoided forcing hierarchy in the faces of their employees, remembering how they'd chafed under the yoke of Shockley and the incompetent bureaucracy of the Fairchild corporate management on the east coast. For example, managers did not garner any special parking privileges at Intel—whoever arrived first in the morning got the best space. The furniture and decoration of managers' cubicles was no different from that of the ordinary staff members' either, and managers also casually and routinely sat down in the company cafeteria amongst other employees to eat lunch and to chat about technology. These practices, while common today, stood in stark contrast to what most other corporations of the time were doing. In fact, much of today's corporate culture of eschewing hierarchy in favor of egalitarianism and meritocracy is heavily influenced by the practices instilled early on at Silicon Valley companies.

Noyce and Moore made a conscientious decision to make memory chips Intel's major business focus from the start, in order to sidestep the distractions of other, tempting technical opportunities. Falling prey to obsession with novel technologies was a common trap for many researchers-turned-businessmen. They didn't want the lack of focus that had befallen Shockley Semiconductor to shape their fate. This decision partially explains the company's initial reluctance to enter the microprocessor business, even after Faggin's success with the 4004. Of course, they weren't stubborn enough hold out forever in the face of such great opportunity.

Though Intel's initial business grew rapidly, the company's memory products began to face intense competition from Japanese firms starting from the late 1970s, and Intel's gross sales and profits margin declined. Even the establishment, through Noyce's intense personal effort, of the Semiconductor Industry Association, a lobbying group that sought to limit Japanese imports to the United States, effected little relief. Eventually, painfully, Intel was forced to undergo sizable layoffs. The company needed a new direction and a new leader.

Moore had taken over from Noyce as president in 1975 and as CEO in 1979, and though he served well, over time, it became clear that someone more leather-skinned was needed: a cold-blooded and iron-fisted man who could hold the company and all its employees ruthlessly accountable while still being the technical and business visionary. At Intel, as it happened, there was just one such man: Andy Grove.

Andy Grove was born Andras Grof in Budapest. A Hungarian Jew, he survived the Holocaust, Soviet-satellite Communist rule, and the 1956 uprising. Still in his youth, he fled to the United States, settling first in New York, where he earned a bachelor's degree from City College, then moving on to Berkeley, where he received his PhD in chemical engineering. From Berkeley, he joined Fairchild, and there, he made a name for himself: he was capable, intelligent, and hard-driven, and he was highly regarded by both Noyce and Moore. In fact, Grove was so well regarded by the two that he became the first person to join them at their new company, making him Intel's third employee ever.

Grove proved very different in his managerial style from Noyce and Moore. He ran a tight ship and was a strict and detail-oriented disciplinarian. Demanding much from himself and his subordinates, he could be tough and ruthless. Grove required everyone to report to work before 8:00 AM. If anyone was late, he or she would be fined, and even Grove, the company president himself, was not exempted. As the marketplace grew ever more competitive, this hardline discipline became the new name of the game.

Though Grove's methods probably would not have been implemented by the company's founders, they worked. Grove spent almost twenty years lead-

Andy Grove (left), Noyce (center) and Moore (right). Courtesy of Intel Corp.

ing Intel, retiring in 1998. During his time in office, Intel's market capitalization grew fifty-fold from $4 billion to almost $200 billion, and the company morphed from a pure semiconductor chip manufacturer to an electronics systems empire with the deepest knowledge of computer architecture. Influenced by his childhood experience as a Jewish boy growing up in hardship under communist rule, Grove tended to think in terms of worst-case scenarios, never allowing complacency to creep in. His motto was "Only the paranoid survive," and odd though it was, the company's long-term success proved the effectiveness of Grove's leadership.

Grove led Intel for years after Noyce and Moore ceded control of their company. Noyce fully enjoyed the life he lived: he took up gardening, singing, flying, scuba diving, and reading Hemingway. Noyce also donated generously to his alma mater, Grinnell College, and to a foundation he started that was devoted to improving public education in math and science. He died in 1990 at the relatively young age of sixty-two, suffering a heart attack while swimming at his home. In his final years, he established SEMATECH (Semiconductor Manufacturing Technology), a unique, nonprofit research cooperative jointly funded by many U.S. companies to develop "precompetitive" semiconductor manufacturing technologies and equipment and to share the use of the results. This joint venture pooled resources from chip producers, chip manufacturing equipment companies, material suppliers, and research organizations. Working together, it aimed to sustain the United States' lead in semiconductor technology against global competition.

Moore, too, was an extremely generous philanthropist. While the media praised Moore as a great technologist and businessman, he modestly called himself "an accidental entrepreneur," and that may have been his most accurate portrayal. In 2001, Moore unconditionally donated $600 million to Cal Tech, at the time the largest donation by any individual ever to an educational institution. In 2005, he set up another foundation with an endowment of almost $6 billion to specialize in environmental protection research. Thanks to "Moore's Law," even today his name continues to pop into the public awareness. Not just in the *Physical Review*, either—he's often mentioned in the *Wall Street Journal*.

TURNING SILICON INTO GOLD

Over the past fifty years, countless numbers of great minds have made contributions to further the growth of chip technology, along the way leading mankind in harnessing the power of electrons. Their efforts also helped to create great fortunes for many people during the process. In fact, this era of Silicon Valley represents probably the greatest sustained period of wealth creation in all of human history. The alchemists of the Dark Ages never got far when they tried to transmute aluminum and lead, but in our time, scientists and businessmen have indeed turned silicon into gold.

In the late 1940s and early 1950s, the area that is today known as Silicon Valley was mostly peaceful and beautiful farm land, dotted with orange orchards, apricot groves, and native oak trees scattered along rolling hills. With the founding by Frederick Terman in 1951 of the Stanford Industrial Park, which pioneered the high tech industrial park concept, some industries began to move in, such as Lockheed, or were founded by local talents, such as Varian Associates and Hewlett-Packard. Even Lee De Forest, the inventor of the vacuum triode, at one time set up his company in Palo Alto. But the region was still mostly sleepy and quiet until 1955, when Shockley started his transistor company and began to recruit talent from all over the country. This would eventually change the region forever. Through the rapid growth period and the spin-off era that birthed Fairchild, Intel, and others, the area transformed itself into the most important hub of high technology in the entire world. Tremendous amounts of wealth have been created as a result, and Silicon Valley is consistently ranked among the most affluent regions in the nation. But how was this wealth created and amassed?

The best example is the case of Moore. As a PhD from Cal Tech with two years of working experience, he was hired by Shockley in 1955 with a starting salary equivalent to $65,000 in today's money, which was fairly unremarkable. When the Traitorous Eight left to found Fairchild Semiconductor in 1957, each man contributed $500 as a token ante. With their effort, Fairchild became a leader in the rising semiconductor industry. Both revenue and profit grew rapidly, and with that, the valuation of the company increased. Two years later, when their shares were bought out, each man's initial $500 investment had turned into $250,000, or approximately $2.0 million in 2012 dollars. This value was created by the men's immense talent, uncommon knowledge, and total dedication. It was also aided by the right environment that allowed such opportunities to blossom.

After leaving Fairchild in 1968, Moore and Noyce invested, curiously enough, the equivalent of their payouts from Fairchild to start Intel. As Intel's business grew, their share of the equity in the company came to be worth billions. Leveraging an initial investment of $500 into billions of dollars is truly a remarkable success story in free enterprise. To think deeper, the origin of the wealth is really derived from the impact of the products from the company which provided enormous value for the users and the society as a whole. When it was founded in 1968, Intel employed three people and had neither products nor revenue. But every time anyone bought a PC or a digital product in subsequent years, very likely they all paid a little toward Noyce and Moore's contribution to make the chip an extremely valuable tool for enhancing productivity in their jobs and personal lives. By 2011, total employment at Intel had reached nearly 100,000 and its revenues totaled over $55 billion, with a net profit of over $16 billion per year.

There are countless successful high tech companies headquartered in Silicon Valley, largely because of the innovative culture and effective start-up infrastructure that had been nourished and built up through years of success of the electronics industry. Many of the founders of these successful companies gained great wealth, just like Moore, and this wealth also trickled down to a wider circle of executives and key technologists through stock options and bonuses. In addition to the technical entrepreneurs who have created great fortunes for themselves, many far-sighted venture capitalists also struck jackpots. People like Rock and Kleiner skillfully leveraged investments and sound business advice into huge sums of capital by helping talented entrepreneurs launch successful high-tech companies. Silicon Valley's infrastructure for new business creation was gradually perfected over time, and in addition to technical entrepreneurs and venture capitalists, lawyers, accountants, headhunters, and special consultants found their niches. All of these elements coalesced into a continuous network that fueled the creation of new high-tech industries.

When the growth of new semiconductor companies began to slow as chip technology become more mature, this new business creation machine pivoted, turning its attention to attractive, emerging opportunities outside the realm of semiconductors, such as computer systems, software, the Internet, social networking, biotechnology, even electric cars. Many entrepreneurs with novel ideas were attracted to the Silicon Valley and its fertile soil for growth. Companies such as Apple Computer, Yahoo, Oracle, Google, eBay, Facebook, Genentech, Tesla, and many others all owe their origins to the combination of their founders' original entrepreneurial spirits and ideas and the power of the business creation infrastructure in Silicon Valley. Two top-tier universities in the region, Stanford University and UC-Berkeley, also played an essential role as the intellectual core of the Valley, providing a continuously renewable source of new ideas and talent with a global reach.

19

LEDS, FIBER OPTICS, AND LIQUID CRYSTAL DISPLAYS

LUMINESCENT SEMICONDUCTORS

Silicon chip technology may occupy the center stage in the modern electronics industry, but chips do not fuel industry growth alone. Over the past fifty years, important hardware technologies such as light-emitting diodes (LEDs), semiconductor lasers, fiber-optic communication systems, liquid crystal displays (LCDs), CD/DVD/Blu-ray discs, mass storage magnetic hard disks, and advanced electronic component packaging technologies have all contributed to the spectacular success of the information age. Some of these technologies related to the generation, detection, and manipulation of light, or photons—a field collectively known as *photonics*—have supplemented electronics in many important ways.

Silicon is an ideal multifunctional semiconductor. However, silicon does lack one important property: it cannot be made to emit light, at least in its basic form. When electrons and holes recombine in silicon, the released energy is converted to kinetic energy in the form of atomic lattice vibrations. These movements are essentially a form of heat. That silicon reacts in such a manner was precisely predicted by theories of solid-state physics. In applying the same theories, it was predicted that in some other types of semiconductors, when electrons and holes recombined, the energy released *could* be turned into emitted light.

Materials like silicon and germanium are known as elemental semiconductors because they are formed of only one type of element in group IV of the periodic table. There is another large class of semiconductors formed through the combination of column III and V elements, and these are known as III-V compounds. One example of these III-V compounds would be gallium arsenide

	III	IV	V	VI
	5 B	6 C	7 N	8 O
	13 Al	14 Si	15 P	16 S
30 Zn	31 Ga	32 Ge	33 As	34 Se
48 Cd	49 In	50 Sn	51 Sb	52 Te
80 Hg	81 Tl	82 Pb	83 Bi	84 Po

Portion of the Periodic Table showing column III and column V elements. Derek Cheung

(GaAs), made by the combination of gallium and arsenic. Theories predicted that most III-V compound semiconductors could be made to emit light—in fact, based on these compounds' energy band gaps, even the color of the light they might emit could be predicted! But as exciting as this prospect seemed, there was a problem: at no point in known history had any of these III-V compounds ever existed in nature.

Armed with the knowledge of the potential of these III-V compound semiconductors, scientists began to attempt to synthesize them artificially. In 1953, a research group at Siemens, under the leadership of an outstanding physicist, Heinrich Welker, became the first to break through, successfully synthesizing single crystals of indium antimonide (InSb) and gallium arsenide (GaAs). Measured results proved that these artificially synthesized III-V semiconductors behaved exactly as predicted by theory. This was a stunning achievement, and Welker's creations led to a series of other breakthroughs.

In the case of InSb, its energy gap—a critical parameter to the behavior of each unique semiconductor—was so small that it quickly proved to be unsuitable for most applications. The properties of gallium arsenide, on the other hand, were very similar to those of silicon but with two important advantages. First, the electron "mobility" in GaAs—the speed at which its electrons move under the influence of an electric field—is much faster than it is in silicon. This made GaAs an ex-

Heinrich Welker. Siemens Corporate Archives, Munich

cellent material for ultra-fast electronic chip applications, such as in wireless communication and radar systems. In fact GaAs chips are used today in all mobile phones as the power amplifier for transmitting wireless signals. The second difference, and the still more intriguing one, was that GaAs could indeed emit light. Excess electrons and holes could be electrically "injected" into a GaAs p-n junction by conducting current through it. When these excess electrons and holes recombine within the material, this recombinant energy was converted into light energy and emitted, thus giving rise to the name light-emitting diode, or LED.

The wavelength of this emitted light is determined by the energy gap of the semiconductor. For gallium arsenide, the emitted light falls in the infrared region of the spectrum, which cannot be seen by the naked eye. However, replacing a portion of the arsenic atoms in GaAs with column V phosphorus atoms would widen the semiconductor's energy gap. This shortened the wavelength of the light it emitted, moving it from infrared into the realm of visible red light.

In 1962, researcher Nick Holonyak Jr. from GE Labs became the first to demonstrate a functional red LED, which was made from gallium-arsenide-phosphide alloy. In the early 1970s, red LEDs began to be commercially mass produced and were used extensively in calculators and electronic watch displays. The red color was bright and pleasant to the eye, and the displays appeared futuristic. A sizable commercial LED market soon emerged. Meanwhile, LED research continued. In addition to making the devices brighter, a significant effort was extended to bring LEDs into still more segments of the visible light spectrum. In the early 1970s, LEDs made from gallium phosphide (GaP), a semiconductor with a very wide energy band gap, were developed. These GaP LEDs could be engineered to emit a range of colors, including red, amber, and green. However, GaP could not be made into emitting blue light, which was a holy grail for wide applications of LEDs.

Why was the blue LED so sought after? It's because by combining red, green, and blue light in varying quantities, it is possible to produce a full spectrum of saturated color, including white. This is why all front-projection TVs of the 1970s and 1980s had three bulbs positioned in front in shades of red, green, and blue. This is also why old, adjustable, color CRT TVs had knobs like "hue" to change the richness of these colors. Without blue, only a limited number of colors could be produced, rendering LEDs unable to adequately display the natural world.

As it turned out, creating a blue LED proved to be an extremely difficult technical challenge. Moving from red to green to blue in the spectrum, wavelengths of light become shorter; correspondingly, the required energy band gap of the semiconducting material grows larger. The value required to create blue light is so high that materials with the required energy gap behave almost like insulators rather than semiconductors. Though creating green LEDs was straightforward once gallium phosphide was developed, it took almost two decades of intense research—including many dead ends at such storied research institutions as RCA Labs and others worldwide—before a remarkable researcher in Japan named Shuji Nakamura solved the puzzle.

One member of the family of III-V compounds that possessed an energy band gap large enough to theoretically be able to produce blue light was GaN (gallium nitride), along with its related alloys. However, for many years, researchers were only able to produce n-type GaN, but not p-type. This made it impossible to forge a p-n junction in this material. The breakthrough came in 1993, when Nakamura demonstrated a bright blue LED built of indium gallium nitride (InGaN), which he synthesized with a unique chemical vapor deposition technique. Through many years of hard, solitary work, he discovered the subtle role played by the hydrogen atom in allowing the creation of both n-type and p-type semiconductors with this material, unlocking the ability to create in InGaN a p-n junction, the key building block of an efficient blue LED.

The development of the blue LED was a major victory for semiconductor materials technology. Surprisingly, Nakamura's employer, the Nichia Corporation, was a traditional chemical company specializing in florescent materials for CRT screens and light bulbs, and it was not active in LED research or the semiconductor business. Nakamura was initially allowed by management to work on his pet project only because of his relentless insistence; however, management support was later withdrawn, forcing Nakamura to work on the research on his own time and with his own funds. Yet, after Nakamura's success, Nichia claimed ownership of the technology, offering Nakamura in 1993 a token award of ¥20,000, the rough equivalent of $180 at the time.

Despite the unfairness of this act, in the quiet, buttoned-up world of Japanese industry, it is strikingly unusual to speak out against an employer. However, Nakamura was a maverick. He felt that Japanese corporations often employed objectionable practices that were unfair to workers, especially true contributors such as Fujio Masuoka of Toshiba who pioneered the Flash memory technology, but was never recognized and rewarded. Unlike other Japanese engineers, Nakamura wasn't afraid to speak his mind. He wrote many sharply worded criticisms about this matter in the aftermath, a tactic deeply contrary to the tradition of Japanese business culture but championed by many engineers who possessed their own pent-up frustrations with overbearing employers. In fact, Nakamura even went so far as to sue Nichia, and surprisingly, he won, receiving an initial award for the sum of ¥20,000,000,000—one million times what Nichia had tried to pay him off with! The company countersued, however, and in the end, the two sides reached a settlement in which Nakamura accepted a payment of $9 million. Though this was a sizeable figure—larger, in fact, than any settlement any Japanese company had ever previously paid to one of its own employees—rights to the blue LED made the firm substantially more money than that in due time. Nichia did enjoy this market success, but Nakamura received the accolades and remuneration he deserved, and he contentedly settled into a career in academia in the United States to continue his research.

The introduction of the blue LED enabled the development of ultra-large, bright, full-color displays. Nowadays, these screens are ubiquitous in sports stadiums and crowded commercial centers such as the Las Vegas Strip or Times Square. But the largest potential impact of the LED is not simply in entertainment. In the coming years, mankind will have a viable replacement to Edison's incandescent bulb: the white LED for lighting.

One way to produce a white LED is to mix lights from red, green, and blue LEDs; however, this approach is very expensive. An ingenious and more economical way is to coat the blue LED with a special fluorescent material, which converts a part of the original blue light into a broad-spectrum yellowish light. When the converted yellow light commingles again with the remaining, unconverted blue light, it produces a soft and bright white light. Nowadays, almost all commercial white LEDs employ this design.

Compared to incandescent light, LEDs and their accessory electronics are relatively expensive to produce. However, LEDs convert electrical energy to visible light directly via quantum mechanical principles, which is theoretically over ten times more efficient and generates much less wasted heat. Additionally, the useable life of LEDs can reach 50,000 hours or higher, some twenty-five to thirty times longer than similarly rated incandescent bulbs. Thus, despite their higher price, white LEDs are already finding commercial utility in flashlights, street illumination, traffic control lights, automobile headlights and taillights, aircraft cabin illumination systems, and television backlighting systems. As their price continues to drop and performance continues to improve, LED lighting finds itself on the verge of breaking into wide-scale home use, as evidenced by the evolving lighting inventory at a typical local hardware store. With continued investment in production technology, the price of LEDs is expected to continue to fall, and in the not-too-distant future, high efficiency LED lights will replace incandescent light bulbs altogether, playing a major role in energy conservation.

SEMICONDUCTOR LASERS

The principle of the laser was first proposed after World War II by Charles Townes. Like many other scientists of his time, Townes spent several years working at Bell Labs before he was assigned to work on military technologies during the war. However unlike many others, he decided to settle in academia after the war and joined the faculty at Columbia University. In 1951, Townes conceived of a unique concept to amplify microwave signals using energized ammonia gas as a medium. His innovation was a clever implementation and extension of a theoretical concept in quantum mechanics known as *stimulated emission*, a postulate first advanced by Einstein. This signal amplification principle was totally different from the electronic amplification principle of vacuum tubes or transistors. By 1953, Townes and his collaborator designed and demonstrated a novel microwave amplifier based on this concept. The sensitivity of the device was unprecedented, and it was quickly embraced by radio astronomers to detect the faintest celestial bodies in the universe. Townes named his device the MASER, an acronym of microwave amplification by stimulated emission of radiation.

In 1958, Townes jointly published a seminal paper with another physicist, his brother-in-law Arthur Schawlow, establishing the theoretical ground that the principle of the maser could be extended from the realm of microwave frequencies into the spectrum of visible light. This prediction set off a race among scientists across the globe to be the first to physically demonstrate that concept. The race was won by Theodore Maiman of Hughes Research Lab in California. In 1960, Maiman employed a ruby crystal in a device that emitted a coherent red light.

Maiman's device was named the LASER, similar to the maser acronym except that the "M" for microwave was replaced with an "L" for light. In 1964, Townes, along with two Soviet physicists who had also independently made major conceptual contributions to the laser, received the Nobel Prize in Physics.

Within a short period of time after Maiman's demonstration, many research groups across the world reported successes in realizing a dazzling variety of lasers using different media, including gases, liquids, and solid crystals. These demonstrations covered a wide range of output light colors and power levels. What makes laser light distinct from other sources of light is its "coherence." Laser light is pure in terms of its wavelength, and therefore, in its color. In addition, the peaks and valleys of the electric and magnetic field oscillation of the travelling laser beam all stack atop one another in unison—that is, they are "coherent." Collimated columns of coherent laser beams can travel over long distances without diffusing, as seen in outdoor laser light shows. More importantly, laser beams can be focused onto incredibly small spots with highly concentrated light intensity.

As more and more variants of laser devices were created, it became clear that the most compact laser devices should be based on semiconductors, and in 1962, researchers at GE and IBM announced the successful demonstration of a semiconductor laser based on a design variation of gallium arsenide LED. Unfortunately, this laser needed to be cooled down to -196°C to function, a temperature so low that its operation was totally impractical. Nonetheless, its demonstration proved the fundamental feasibility of the concept. The next step was to make a semiconductor laser practical.

In order to create the conditions needed for stimulated emission in semiconductors, the concentration of both excess electrons and holes must reach a very high threshold value, much higher than the operating conditions required in a normal LED. However, excess electrons and holes near a p-n junction tend to diffuse away from each another, so it is difficult to keep them spatially close together for long enough a time to reach the threshold value necessary for stimulated light emission to take place. At the initial demonstration in 1962, researchers at GE and IBM relied on brute force to solve the problem: they provided a gross oversupply of excess carriers by injecting a very high current across the p-n junction. But under such conditions, the device would overheat and burn out easily. This is why the first working GaAs laser had to be immersed in liquid nitrogen at -196°C—to overcome these problems.

The key challenge to operating a practical semiconductor laser without cooling thus became solving the problem of keeping excess electrons and holes from spreading away from each other near the p-n junction. A clever conceptual solution first came from German-American physicist Herb Kroemer. In 1956, before the demonstration of any lasers, Kroemer proposed the concept of a heterojunction, a junction between two different semiconductor materials with different energy band gaps. Kroemer predicted that the adoption of the heterojunction concept could lead to major performance improvement in all semiconductor devices. In 1963, after the initial demonstration of the GaAs semiconductor laser, he recognized and proposed that if a thin layer of material with a smaller energy band gap were sandwiched between two materials with larger band gaps, the excess electrons and holes in the middle layer would be confined, like particles locked up in a box. If this concept could be incorporated near a p-n junction, then the electric current necessary for the device to reach the threshold needed for stimulated light emission would be much reduced.

Kroemer's solution was simple and elegant, making the new challenge one of developing the materials technology necessary to build such heterojunction structures. This challenge was addressed by the development of material deposition techniques by which thin layers of single crystal materials, such as GaAs and their alloys, were deposited atop each other one atomic layer at a time. Thanks to such precise levels of process control, engineers were able to design and fabricate sophisticated structures, such as heterojunctions, at the atomic level. By the late 1960s, high quality heterojunctions between GaAs and its larger band gap alloy AlGaAs (aluminum gallium arsenide) were successfully demonstrated in several laboratories around the world. Not long afterwards, in the spring of 1970, Zhores Alferov of the Soviet Union and researchers at Bell Labs were almost simultaneously able to demonstrate a heterojunction-based GaAs laser that could operate continuously at room temperature without any active cooling.

The heterojunction and its application in the semiconductor laser came to serve as the core technology behind such information age products as CD/DVD/Blu-ray, bar code scanners, and fiber-optic communication systems. Interestingly, when the laser concept was first conceived of, no one had any specific applications in mind for its use. Even Townes himself confessed that he had only a very vague idea that the laser might somehow be employed in communications technology. Thus, the laser was initially an invention begging for applications, a hammer looking for a nail! Of course, this period did not last long.

In 2000, Kroemer and Alferov were jointly awarded the Nobel Prize in Physics, together with Jack Kilby. By that time, CDs were already in widespread use, but DVDs were just entering the public consciousness, and Blu-rays were still under research and development. What is the difference between these three similar-looking discs? The answer lies in the wavelength differences of the laser beam that reads and encodes their data.

For optical systems using laser beams, the smallest focused spot size is ultimately limited by the wavelength of the laser: the shorter the wavelength, the smaller the spot size. When they were first introduced, CD systems used GaAs lasers, which emitted light in the infrared spectrum. A first-generation CD could hold about 650 megabytes (MB) of data, which translated into about 74 minutes of uninterrupted CD-quality audio. (Interestingly, this standard was heavily influenced by Sony's founder Akio Morita, who insisted that Beethoven's entire ninth symphony must fit onto one CD and not be split in two.) As red lasers replaced GaAs lasers and more precise motion control techniques and error correction algorithms were added, CDs were replaced by DVDs, which boasted much higher data storage capability, up to 9.4 GB for a two-layer configuration, which is the equivalent of fifteen CDs. Instead of just music, full length movies could be recorded on DVD discs. As blue lasers became available and replaced red lasers, the DVD was supplanted by the Blu-ray discs, which can store an astonishing fifty gigabytes of data on a two-layer configuration. Now, ultra-high definition, full-length 3-D movies can be recorded onto and read from a single Blu-ray disc of the same size as a first-generation CD, but with seventy-six times the capacity.

FIBER-OPTIC COMMUNICATIONS

When Bell Labs first began serious research into semiconductor lasers, their intention was not for CD, but for utilizing the coherent beams for optical communication. Frequencies of visible electromagnetic waves are about 100,000 times higher than the microwave frequencies used in cell phones; therefore, light waves could in theory be used as carrier waves to transmit enormous amounts of signal information. As one of Bell Lab's long-term goals had always been a videophone—an application that required broadband data rate—optical communication is always an attractive research candidate.

An obvious and simple way to communicate using light is to transmit through open space—for instance, by shining a flashlight at a friend, clicking the light on and off as in Morse code. But while such a method is fine for line-of-sight communication over short distances, over longer distances, sources of atmospheric interference such as fog and rain can attenuate light and reduce its range. Dust and other airborne particles can also scatter light, and temperature differences in the air can refract, or bend, light paths. These fundamental limitations, to say nothing of the need to accurately align and continuously realign the optical source and receivers, are serious and not easy to overcome. But what if there were a different way to send the light beams? Perhaps if light could be guided and propagated through a glass tube, just as electric current flows through copper wire or water flows through a pipe, optical communications could then be a real possibility.

In the mid-nineteenth century, French scientists discovered that when shining light on a fountain, some of the light would become trapped in the columns of water. Later, an English scientist took that discovery one step further and determined that light could be guided and even bent along a curved water flow. Still other scientists demonstrated these same light-trapping and light-guiding phenomena using a curved rod of clear glass.

Channeling light down a glass rod proved a relatively simple feat, but glass rods were too rigid, fragile, and bulky to be widely implemented. In the art of glass blowing, there was a traditional technique to pull hair-thin strands of glass fibers from a melt. The diameter of the glass fiber could be controlled by the speed of pulling and the viscosity of the melt. Unlike glass rods, glass fibers are relatively mechanically pliable, and they can be bent to a certain curvature without breaking.

In the early 1950s, engineers developed a way to bundle glass fibers together to form a rope-like structure that was flexible and could be slightly bent into different curved shapes. By polishing both ends of the fiber bundle, images projected on one end could be seen from the other. These flexible glass rods found an application in looking into physical spaces that human eyes could not easily access. The most valuable such application was in looking into internal parts of the human body, such as the stomach, and soon a medical device known as the endoscope was developed to do just that. But even though the device was useful, it was expensive, and the optical image it offered lacked contrast and clarity due to the excessive attenuation of the light propagating through the glass fiber. This high rate of light loss represented a key technical issue to be solved, not only to improve performance of short-range applications, but also for the ultimate feasibility of long distance optical communication.

Researchers began to turn their attention to understanding the fundamental causes of light attenuation in optical fibers. It soon became clear that there were two major causes. The first was reflection loss. As light propagates along a fiber, it bounces forward at an angle, reflecting multiple times at the interface with air and losing light energy each time. Worse yet, these losses are multiplicative. For example, with a loss of just 1 percent after any given reflection, after 1,000 reflections, only 0.004 percent of the original light remains. The second loss mechanism was absorption of the light by the glass material itself. Even though these losses were relatively minor over short distances, over long distances, their effects were exacerbated.

Tackling these issues one at a time, researchers began to first examine methods to reduce signal losses due to reflection. The crucial breakthrough came from Lawrence Curtiss, an undergraduate research assistant at the University of Michigan. At the time, he was working under the guidance of his professor to improve optical fiber losses in endoscopes. Instead of using highly reflective metal to coat, or "clad," the optical fibers, which was the mainstream research approach at that time, he decided to see what would happen if he replaced the metal cladding with another thin layer of glassy material—one that had a slightly lower refraction index than the glass core fiber itself. The lower refraction index of the outer glass caused the light wave within the core to be reflected at the interface through the "total internal reflection" mechanism, with much reduced loss. The outer glass cladding layer also provided effective protection for the delicate glass fiber core.

The low refraction glass cladding technique elegantly solved the problem of reflection loss and greatly enhanced the performance of endoscopes and other medical instruments. However, there still remained the problem of absorption loss. At the time, the light loss rate in the best fibers was about 10 percent per meter. The effect was multiplicative, too, so after just 40 meters, only 1 percent of the original light signal would be left—99 percent would be absorbed by the glass fiber! With loss rates like that, long distance fiber-optic communications remained a dream, and people were clueless as to what the next step should be. Many large research firms, including Bell Labs, put their optical fiber–based communication research on hold and turned their attention to alternative approaches, such as using hollow metal tubes to guide light. But not everyone in the industry had abandoned the idea.

Charles Kao was born in Shanghai, China, but moved with his family to Hong Kong as a teen and immigrated to England for his college education. He received a PhD while working as a student researcher at the Standard Telecommunications Laboratories, or STL, near London, which was a subsidiary of the telecommunications giant International Telephone & Telegraph. Even though STL was much smaller than Bell Labs, it was an excellent and innovative place to work on cutting-edge communications technology. The young Kao first did research on microwave communication, and later, he became intrigued by optical fibers. Through his hands-on experimentation, Kao became convinced that fiber-optic communications held great potential. Working closely with another engineer at STL, George Hockham, Kao formulated a focused research project aimed at bringing fiber-optic communication to reality.

Employing a highly methodical approach, Kao first designed and built a simple point-to-point fiber-optic digital communications test bed using a simple helium-

Charles Kao. Courtesy of the Chinese University of Hong Kong

neon laser as the signal source, optical fiber as the transmission medium (cut very short, so that optical loss would not be an issue), and a photo-detector as a signal receiver. Kao and Hockham demonstrated that under laboratory conditions, fiber-optic communications systems could achieve astonishing data rates of one gigabit per second or above—that's one billion binary points of information transmitted and received in one second! Convinced of the potential of fiber-optic communications by that first physical demonstration, they began to mathematically model the system performance under a host of different parameters. After extensive analysis, Kao was able to conclude that under realistic parameters expected based on then-current laser power sources and detector sensitivity, a practical and useful broadband fiber-optic communications system *could* be built if the absorption loss in the optical fiber could be reduced to less than 99 percent per kilometer. That is to say, if just 1 percent of an original optical signal could get through one kilometer of fiber optic cable, the system could find practical applications. That 1 percent per kilometer quantitative threshold was the key. But the question remained: was achieving such a benchmark physically possible?

To find the answer, Kao began to gather swatches of optical fibers produced by manufacturers and research institutes around the world. After meticulously and systematically measuring the optical losses of each sample, Kao concluded that losses in optical fibers were primarily due to light absorption by unwanted impurities in the glass and not due to some fundamental property of the glass itself. If the impurity content in glass optical fibers could be reduced, then bringing the signal loss rate below 99 percent per kilometer would be entirely feasible. Just as during the dawn of the era of the semiconductor, the bottleneck lay in materials science and fabrication technology, and it was from those fields that a champion would be needed to find a breakthrough. Kao published his key findings in 1965, but the work only received limited attention from the technical community.

Kao's results pointed out a clear and fruitful path for fiber-optic research. So, with great zeal, he began to travel to the leading labs and research institutes around the world to promote his views. Some scientists were intrigued; others viewed him as a crackpot. At Bell Labs, particularly, his ideas were looked upon with significant levels of doubt and scorn—due, perhaps, to institutional bias by employees who felt certain that a young researcher working for a small lab could not possibly have found the answers that had eluded their best men. Nonetheless, Kao's results energized the optical fiber research community, and with a clear target now in view, many companies resumed their efforts to develop low-loss fibers.

At first, most companies concentrated on trying to purify their glass material in its solid form. However, after two years of development, progress remained minimal. In 1970, researchers at the Corning Glass laboratory took a totally different approach to solve the problem: instead of purifying the glass solid or melt, they focused their effort to purify the reactive gases that were used to deposit the fused silica from which the thin fibers were extracted. Using this process, they were successful in producing glass fibers of exceptional purity. However, the refractive index of this highly purified silica proved to be too low to be used with cladding. Corning researchers then developed a technique to precisely increase the index of refraction by adding traces of titanium, or germanium. The process is analogous to the doping process used in the silicon chip industry. The combined two-step process of gas phase purification and impurity doping paid dividends. In 1970 Corning Glass announced their new fibers achieved optical losses of less than 98 percent per kilometer, exceeding the minimum standard set by Kao. Equipped with a winning technological process, the company redoubled its efforts, and within two more years, their glass fiber's loss rate was reduced significantly from 98 percent to 60 percent. By the midpoint of the decade, low-loss optical fiber entered commercial production status, and once it did, the fiber-optic communication industry began to take off like a rocket. Today, light loss in fiber-optic cable can be less than 4 percent per kilometer—which means more than 96 percent of light can travel unimpeded through a 1,000 meter-thick glass.

1970 was a monumental year for fiber-optic communication, as it was not only the year that Corning produced the first low-loss fiber, but also the year Bell Labs demonstrated the first practical semiconductor laser. The convergence of the two technologies ushered in the age of broadband communication. The first commercial optical fiber link went into official use in 1975. Thirteen years later, the first transatlantic fiber-optic cable was completed, and shortly thereafter, nearly the entire world was bound together by broadband optical fibers. This network, like a spider web that covered the whole globe, would become the backbone of the entire Internet.

As technological advances continued in optical fibers, lasers, and high-speed electronic chips that interface with the optical medium, data rates in broadband communications increased exponentially, just like Moore's Law for chips. Additionally, thanks to a technology called wavelength division multiplexing—a technology much like the harmonic multiplexing concept for the telegraph introduced by Elisha Gray and Alexander Graham Bell—it was demonstrated that one strand of optical fiber could simultaneously carry not just one laser transmission, but more than 150 lasers of slightly differing wavelengths at any given time. Such a cable can transmit more than *fifteen trillion* (15,000,000,000,000) binary bits of digital information every second, through distances of up to 7,000 kilometers. These performance levels are mind-boggling, and they are directly responsible for allowing hundreds of millions of users to simultaneously watch streaming videos and make videophone calls with friends in other continents from the comfort of their own homes, at minimal cost. Just imagine our voices and images, which are encoded in ultra-short pulses of light, bounced forward in optical fibers across thousands of miles, under the sea and over the mountains, at nearly the speed of light, to reach the destination. These are not science fiction scenarios; they are happening everywhere, every second.

As with so many outstanding scientists before him, in 2009, Charles Kao received the Nobel Prize in Physics. Kao was always a humble man, and on many occasions, he sincerely and modestly stated that his research was quite simple and nothing extraordinary—he just did the legwork and pointed out the direction to make fiber-optic communications practical. That's fine to say, but during his time, the worldwide optical communication research community was in despair and confusion, without any sense of direction. It was Kao's work that expelled the doubts and defined a clear and productive direction with meaningful quantitative metrics. In a complex labyrinth, Kao pointed out the right path to the exit, and for that, he deserved all the honors he received.

LIQUID CRYSTAL DISPLAYS

The optic nerves that connect the eyes to the brain consist of over one million nerve fibers, each of which is connected to multiple sensor cells on the retina. Constant analysis of visual stimuli signals accounts for a significant part of human brain function. Display technology is therefore the most crucial and effective interface between human users and the computer.

RCA gave commercial CRT televisions their public debut in 1939. For the first sixty-five years of their existence, television displays were dominated by CRT technology. Computer screens, too, employed CRT monitors to visually output data for decades. However, from the beginning of the twenty-first century, liquid crystal displays, or LCDs, began to displace the CRT. In short order after the large screen LCD was introduced, the bulky CRTs that had been familiar sights for generations essentially vanished from both homes and offices. How did this new technology come so rapidly to replace its predecessor?

Though it seems novel, the liquid crystal has over a century of history. In 1888, Austrian scientist Friedrich Reinitzer extracted a derivative of cholesterol from carrot juice while working as a physiochemical researcher at Charles University in Prague. He discovered that this derivative possessed a property unlike anything he had ever previously seen: it appeared to have two melting points. When the substance was first melted into liquid form, the color turned milky, yet at this point the substance exhibited many properties like those of a solid crystal, hence the name *liquid crystal*. Further application of heat allowed the substance to reach a second melting point, at which its color turned clear and its crystalline properties disappeared. This process was reversible and repeatable as the substance was cooled and reheated through multiple cycles. Though Reinitzer discussed these findings with colleagues and publicly presented them, the scientific world, for many years, did not pay much attention to it, and it attracted only a small number of devotees. It was generally viewed as esoteric and without any practical value.

Though it was Reinitzer who discovered liquid crystals, the real pioneer of the first innovative, successful commercial application of the material was George Heilmeier. Over the course of his extensive career, Heilmeier would come to serve as chief technical officer of TI; president of Bellcore, a major spin-off of AT&T; and the director of DARPA, the agency responsible for funding and developing breakthrough technologies for military applications—many of which often find civilian uses as well. At DARPA, Heilmeier made several key decisions that influenced the development of stealth aircraft, a new generation of spy satellites, artificial intelligence, and the foundation of the Internet, among other things. But his true claim to fame stems from his research into liquid crystals.

George Heilmeier and LCD. Courtesy of Hagley Museum and Library

Heilmeier was born in Philadelphia in 1936, and he attended the University of Pennsylvania, where he studied electrical engineering. From school, he joined RCA's research lab in Princeton, New Jersey, and while there, he won a scholarship to pursue a PhD degree at Princeton University during his off hours. Like many other young scientists of the early 1960s, Heilmeier focused his energies on semiconductor research. Silicon chip technology had just taken off, and there were plenty of growth opportunities in the field. As it happened, RCA Laboratories housed one of the most innovative research teams in the industry: it was the first to successfully fabricate a chip using MOS technology in 1962, and it was the first to make a CMOS chip in 1967, both major milestones. Heilmeier, for his part, was an ambitious young man, and he looked forward to working in a technical field that would provide him an opportunity to break out and shine. The semiconductor field seemed to be too crowded.

While in graduate school, Heilmeier foundered about as he sought a topic for his PhD thesis. His supervisor at RCA encouraged him to specialize in organic semiconductors, which was then an obscure field but one that had begun to show promise. Sensing that this might be his chance to make a name for himself at the point of inception of a new technology, Heilmeier switched his focus from semiconductor research toward organic chemistry, with a special emphasis in organic electronic materials. Fueled by a new sense of purpose, Heilmeier completed his PhD in less than two years, and he then rejoined RCA full-time.

The first research project Heilmeier was charged with at RCA after getting his PhD was to develop organic, optoelectronic materials. RCA planned to use these materials to modulate the amplitude of light emitted from lasers in high speed optical communications. Heilmeier accepted the assignment, but what really caught his attention was the work of a senior researcher at the lab, Richard Williams, who was studying various properties of liquid crystal materials. Some of the properties exhibited by these materials aroused Heilmeier's imagination. With his electrical engineering background and an instinct for commercial applications of scientific phenomena, Heilmeier saw great potential for using liquid crystals in optical displays.

A born leader, Heilmeier soon became the head of RCA's liquid crystal research team. He directed the group members to focus all their disparate projects toward one goal: developing an electronically controlled liquid crystal display. In 1964, Heilmeier's group found that when a thin layer of a special type of liquid crystal was sandwiched between two pieces of glass with transparent electrodes, the see-through liquid crystal could be induced to turn milky-white by the application of an electrical field. Suddenly, the prospect for creating a commercial LCD product became very clear. A research field that half a decade earlier seemed an intellectual backwater now appeared to be bursting with product possibility.

Over the next few years, Heilmeier's group improved their ability to control the liquid crystal material and successfully incorporated it into a new type of flat, electronically controlled display devices. Finally, in 1968, the team completed work on a family of LCD sample products, including black-and-white alphanumeric display units whose applications appeared endless. Unlike the CRT, the LCD did not require a vacuum enclosure or high-voltage power supply for a scanning electronic beam. As a result, LCDs were flat, compact, lightweight, low power consuming, and safe. These flat displays were ideal for use in calculators, watches, meter counters, medical instruments, or any other instrument panel that utilized an information display. RCA held a special press conference to announce this important invention, and Heilmeier and his team basked in the glory of their success. Being a visionary, Heilmeier even predicted that large, flat television screens that could be hung flush upon a wall would one day be developed using liquid crystal technology.

Unfortunately, the moment of joy for the team was short-lived. Following the public announcement, Heilmeier approached RCA management to discuss detailed commercialization plans for LCDs. To his utter disappointment, however, RCA management decided not to further pursue LCD technology, despite all the time and funding they'd invested. The reasons they gave were that RCA did not manufacture any products that required this type of simple display, and the LCD technology did not appear to be a creditable challenger to the CRT for adoption in the television, despite Heilmeier's proclamation. The bottom line was that RCA management was simply not interested in investing any more resources into what they viewed as a "non-core" technology.

When it became obvious that LCD had no future at RCA, members of Heilmeier's group began to drift away. One of the key liquid crystal scientists, Wolfgang Helfrich, went to Switzerland to pursue research on the fundamental chemistry of liquid crystal materials. In 1970, Heilmeier himself left RCA as well, going on to become a White House Fellow and a special research assistant to the Secretary of Defense.

While the LCD team was disintegrating at RCA, Sharp, a rising Japanese electronics corporation that had supported Busicom's efforts with Intel for the development of a microprocessor, was looking for a solution for their new portable calculator display. Sharp had been a world leader in transistor-based desktop calculator products since 1964. However, in 1968, TI introduced its revolutionary chip-based pocket calculator, which seriously threatened all other desktop calculator manufacturers. Instead of retrenching, Sharp vowed to fight back with an even more technologically advanced calculator of its own.

The three key technologies Sharp planned to use were low-power CMOS chips for the computing core, a unique chip-on-glass packaging technology to shrink the calculator's size and reduce its weight, and a low-power display for its readout screen. Sharp already had a cooperative agreement in place with Rockwell International for the CMOS chips and a backup option with Intel through Busicom, and the advanced packaging technology had been developed by Sharp in-house. Only the display technology was missing from Sharp's grand plan. There simply didn't seem to be an ideal candidate. Sharp's engineers felt that a red LED display was not a suitable choice for three reasons: first, the LED's power consumption was too high; it would drain the battery. In addition, LEDs required voltage and current levels that were not readily compatible with direct outputs from the CMOS chips. And finally, all the other competitors were already using LEDs. If Sharp did the same, their calculators would lack any major differentiating factor in the market place, and they needed something to make them stand out in the crowded field. There had to be some other display technologies out there. They just didn't know what.

Just when Sharp was searching for the right display technology, Japan's national television broadcaster, NHK, produced a series of special television programs in which it visited many electronics companies around the world and introduced their new technologies. One episode among these featured a visit to RCA Laboratories in which George Heilmeier made an appearance to explain

LCD technology. As it happened, Tomio Wada, the executive at Sharp in charge of calculator displays, happened to see this episode as it aired. Watching the show, he could hardly contain himself, for he was sure that this technology was just what he had been looking for.

In 1969, a team from Sharp eagerly arrived at RCA Laboratories to view these LCD demonstration units with their own eyes. (It was the same year that Busicom visited Intel to jointly develop the CMOS chipset for the new pocket calculator, leading to the development of the first microprocessor.) They liked what they saw, and Sharp immediately decided to try to forge a cooperative arrangement with RCA to jointly develop the LCD and then mass-produce LCD screens. RCA management, however, was not interested in Sharp's collaboration proposal. All they wanted to do was out-license the technology and negotiate patent royalties. To them, the LCD represented a non-core technology; if they could make some money off it, they would do so, but they were not about to devote any more resources to its development.

The delegation returned to Japan disappointed by this rebuttal. Their first choice, a joint development project, was off the table, and the Sharp team subsequently dithered over reaching a strategic decision. But Wada firmly stood his ground, even against voices at the company that told him to find another angle of attack. He reasoned that liquid crystals did not emit light and only reflected natural light. This made them easy on the eyes and kept their power consumption low, which was good for battery life and for ease of integration with CMOS chips. Wada was confident that Sharp engineers, working on their own, could solve the remaining technical hurdles associated with producing a reliable commercial LCD, and he petitioned Sharp's management to accede to RCA's counter-offer and license the existing LCD technology outright.

After much deliberation, Sharp's management approved Wada's proposal, forming a group of over twenty technical staff under Wada's leadership to concentrate on developing LCD technology for calculators. If they doubted that the project would prove successful, they underestimated Wada. Within one short year, Wada's group had synthesized and evaluated over 3,000 different formulations of liquid crystal materials; they had also vastly improved the design of the LCD display. Many advances were made in liquid crystal materials manufacture during this development process, including the accidental discovery by one of the researchers that exposing some liquid crystal material to high electrical fields improved its optical transparency. By the end of 1971, the performance characteristics, reliability, and production technology of LCDs were at last well established, and Sharp owned most of the intellectual property rights it needed to put LCDs into production.

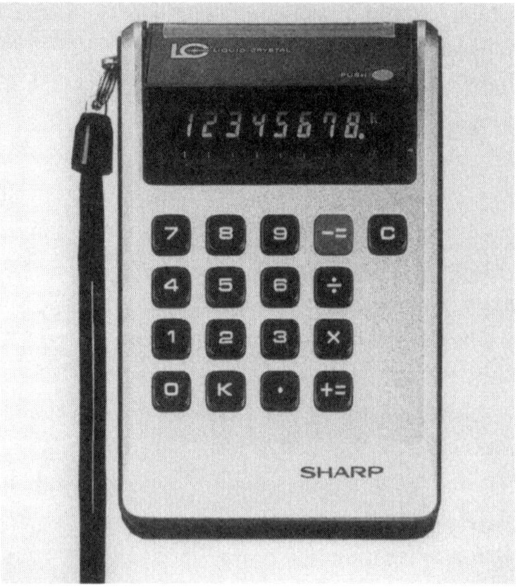

The Sharp ELSI-805 handheld calculator. Courtesy of Sharp Corporation

With the feasibility of LCD fully demonstrated, Sharp launched the final development phase for what they hoped would be their revolutionary pocket calculator. To show the company's total commitment to this product, all Sharp staff members working on this project wore gold colored ID badges, similar to the one worn by the president of the company. Below the names on each badge was printed the number "734," a constant reminder that the product's scheduled date of completion was "'73-04": April, 1973. This level of total team commitment and buy-in did the trick. In early May, 1973, Sharp introduced its new pocket calculator: the ELSI-805. Compared to other calculators on the market, this revolutionary product was 1/12 the size and 1/125 of the weight. One lone AA battery could power the calculator for over one hundred hours of continuous operation. This landmark product brought instant success to Sharp and allowed it to immediately challenge TI's leading position in the pocket calculator market. It also represented the debut of the LCD.

Although Sharp's success put LCD technology on solid footing, the performance level of early LCDs was still not as high as Wada had hoped. These early LCD displays were based on a physical principle called dynamic scattering mode, or DSM. Heilmeier's team first developed DSM while at RCA; they discovered that by applying an electrical field past a threshold value to a liquid crystal mixture, the material affected would strongly scatter light and become opaque. This principle could be used to form visible images. However, the quality of DSM images was barely acceptable: its contrast level was low, and the range of potential viewing angles was highly restricted. Additionally, DSM-based devices showed slow response times to update or refresh, and their power consumption levels, though outstanding relative to other available technologies, were still far higher than expected because they required constant exposure to an electric field. To significantly improve the LCD (and the ELSI-805 along with it), a new approach would have to be developed.

Though Heilmeier's team at RCA had broken up, the Swiss scientist Wolfgang Helfrich had continued his research into LCDs. After returning to Europe, he joined pharmaceutical giant Hoffman-La Roche, and by 1972, he and his coworkers had discovered a new type of liquid crystal material called twisted nematic (TN). The TN LCDs solved most of the problems Wada and Sharp had identified in DSM. TN cells didn't require constant exposure to an electric field to operate. Instead, a single, tiny application of an electric field would coax TN liquid crystals' molecular chains to twist or untwist, making the material appear either transparent or opaque to polarized light for extended periods of time. The response time, contrast ratio, viewing angle, and power consumption of TN LCDs were far superior to their predecessors. Helfrich was not alone in working on TN research. Liquid crystal expert James Fergason, a former employee of Westinghouse, also developed a similar technology and obtained several patents for his invention. There was the eventual, unavoidable patent war. However, this case was short: Hoffman-La Roche, leveraging its massive financial strength, simply made Fergason an offer to buy all his patents that he could not refuse.

Recognizing the importance of the TN breakthrough, Sharp immediately licensed the basic technology and quickly replaced all DSM-based technology in their product lines with the new TN technology. It was the right move, and by 1974, Sharp had established itself as the indisputable global leader in LCD manufacturing.

As TN liquid crystal technology continued to improve, electronics companies began to dream bigger: what if they could go beyond the world of tiny calculator and instrument displays and employ LCDs in televisions and computer screens? That would lead to a new generation of products with ever-more compact size and lower power consumption. It would even indeed be possible to produce a flat television screen that could be hung on a wall. Such a thing had never been done before, but it began to look more and more feasible. Still, questions remained. For instance, how could a full color LCD be manufactured? The answer was to attach checkered color filters of red, green, and blue over each liquid crystal pixel—a combination of those three shades could be used to create any hue needed. A much more difficult technical challenge would be precisely controlling the video signal applied to each pixel for the viewing length of each individual frame. However, this, too, was solved, and in 1983, Japan's Seiko Watch Corporation successfully demonstrated a small but very high resolution TN-based LCD that was packaged as a working, wristwatch-sized television. Great publicity was generated when this gadget was prominently featured in a James Bond movie, *Octopussy*, released that year.

Seiko's display employed a 2-D switching array fabricated onto a single-crystal silicon chip that was overlaid and electrically connected to a matched LCD matrix, pixel by pixel—31,920 in all. However, this approach was clearly impractical for larger LCDs, since the size of the silicon switch array had to match the size of the liquid crystal display itself, and the cost of such a large area silicon chip would be prohibitively high for any mass market product. Still, the existence of a solution, however flawed, both made clear that a device was possible and also got people thinking: what other, more economically viable methods might there be to create large size electronic switch arrays that could be matched to the large size of practical display screens?

Back when the OPEC oil crisis first hit in the early 1970s, many institutions began to develop low-priced amorphous silicon solar cells for which the silicon thin film was directly deposited on large sheets of glass or other inexpensive substrates. Structurally, amorphous silicon is different from single-crystal silicon—the silicon atoms are connected in a random network with no long-range order. Additionally, the electronic properties of amorphous silicon are significantly inferior to those of their single-crystal counterpart: their solar energy conversion efficiency was in the range of 3 to 5 percent, compared to over 10 percent for single-crystal silicon cells at the time. However, because amorphous silicon solar cells were so much cheaper to produce, they nonetheless managed to carve out a niche in the marketplace.

For a long time, the dream of semiconductor materials researchers had been to develop high-performance transistors over large area, thin-film silicon material deposited on an inexpensive substrate such as glass, rather than on bulk single-crystal wafers. If successful, this approach would not only significantly reduce the cost of silicon chips, it would also make it far easier and cheaper to match chip size to screen size for flat-panel LCDs. Unfortunately, for a long time this dream could not be realized, simply because the electronic properties of the silicon thin films were not adequate to fabricate high performance p-n junctions for effective switching functions.

Through the late 1970s and early 1980s, many research groups around the world continued their research to refine and optimize the properties of thin-film amorphous silicon material, with the goal to enhance solar energy conversion efficiency. One of the leading research teams was the group at the University of Dundee in Scotland. The main breakthrough to come out of the Dundee group was the value of adding hydrogen atoms to amorphous silicon during the thin-film deposition process. "Hydrogenated" amorphous silicon, they discovered, produced much better p-n junctions, which were essential to improving solar energy conversion efficiency. As an extension of this research and with no specific application in mind, the Dundee group also fabricated and demonstrated a hydrogenated amorphous silicon field-effect transistor (FET) with a structure quite similar to Shockley's original 1938 design.

As it happened, Sharp, the corporate champion of the LCD, was at that time also one of the world's major producers of thin-film solar cells. In 1982, the research team from Dundee visited Sharp to exchange technical information. At the end of their discussion, the Dundee group casually mentioned their successful experiments with hydrogenated thin-film FETs. Even though the performance of these FETs was still far inferior to similar devices made with single-crystal silicon chips, they were good enough to serve as pixel switches in LCDs. Quietly, the Sharp engineers marveled; it was all they could do to choke back their excitement. This, it turned out, was the last missing link to making the dream of an LCD television a reality!

Thanks to TN liquid crystal material and thin-film amorphous silicon FET switches that could now be deposited on a large, inexpensive glass substrate, the core technologies for a liquid crystal television that could be mass-produced were suddenly in place. After several more years of hard development work, Sharp successfully introduced the first 14-inch, full color, full motion LCD screen in 1988. This flat and power-efficient screen, with size and video performance comparable to color CRTs, enabled a new generation of portable computers. Previously, efforts at "portable" computers used CRT displays, making them bulky, heavy, and luggage-like. Though user demand for portable computers was high, the product never took off until color LCDs took the stage. LCD displays quickly proved to be the perfect solution for portable computers, and despite their comparatively higher price tag, sales of LCD-equipped portable computers—now called "laptops" thanks to their diminished size and weight—skyrocketed. Consumers were happy, and manufacturers were too, especially since the LCD faced no undercutting competition from other display technologies and there were just a few Japanese manufacturers in the marketplace, led by Sharp. These companies made a significant profit on LCD products and they were then able to reinvest that money into improving technology and enhancing their manufacturing yield, cementing the continuous improvement of the product and the growth of their business.

As more aggressive and capital-rich companies in Asia entered the growing market, LCD sizes grew bigger and their price kept on dropping. Eventually, LCDs completely replaced CRTs, not only in laptops, but also in desktop computers, point-of-sale terminals, and even most television sets. By 2006, affordable, large screen, high-definition, flat-panel TVs began entering people's homes. The futuristic dream of having a large television screen hanging flush on the wall had at last become a reality.

The next generations of consumer electronics technology—tablet computing devices, for instance—all rely on LCD screens with touch-sensitive sensors as their user interface. The performance level and features of these products have gone far beyond what Heilmeier could have ever expected when he was floundering around for a research topic!

20

THE INFORMATION AGE AND BEYOND

PUTTING IT ALL TOGETHER

The collective effort by technologists in recent years has created a broad range of technological building blocks, including multifunctional Systems-On-a-Chip (SOCs), multi-gigabit memory chips, high resolution CCD imagers, bright LEDs, sensitive and high resolution touch screen LCDs, high energy-density rechargeable lithium-ion batteries, miniaturized antennae, high efficiency GaAs microwave amplifiers, and advanced packaging technology that can fit all these components reliably into a tiny package and then efficiently dissipate the heat. To a creative system product designer, this breadth of choices is like walking into a candy shop for a little kid. The challenge is to conceptualize a winning end product with the right combination of features, performance level and cost, and then pick and choose the best building blocks from different sources to implement the optimum solution to create value for the end users. In some cases, these standard building blocks need to be further customized to create product differentiation. In addition to hardware technology, software technology also plays a major role, with options ranging from efficient, stable, secure, and user-friendly system software that allows the platform to flawlessly perform specific, complex tasks, to applications software that offers a myriad of diverse applications to end users.

The most successful integrated product is pioneered by the iPhone. Introduced in 2007, it inspired an entire generation of smartphones and tablets. Thanks to the ever more powerful SOCs and other enabling components, the iPhone performs well as a mobile telephone, a personalized radio, a television, a recorder and music player, a flashlight, a computer, and a GPS receiver, all in one tiny package. In addition, it serves as a web-surfing terminal, a digital camera and video recorder, a musical instrument, an eBook reader, an electronic game console, a finger print identifier, and an interface for over 1,000,000 other applications through downloadable "Apps." Except for radar, almost every electronic function discussed in this book has been incorporated into this tiny, revolutionary product. Yet even with all the incredible functionality and system complexity, the product is still user-friendly, reliable, and affordable to the masses. The iPhone and its derivatives has proven to be a key personal appliance that is catapulting the information age that started in the mid-1970s to a new and more impactful phase; it has even changed the social behavior of a vast population around the world in a very short time period.

The sustained, exponential growth of silicon chip technology has provided the key driving force for the rapid creation and evolution of all types of electronic products. Every six to twelve months, new models of products are introduced that are faster, more versatile, and cheaper than the ones that came before. This product trend fuels market expansion, which in turn creates still more demand for chips, a virtuous cycle. In 2010, total semiconductor chip revenue worldwide was close to $300 billion, and chip fabrication and related support industries accounted directly for approximately ten million jobs. But that's just the chips! Not even components such as LEDs, LCDs, fiber optics, and batteries are included in the figure. Revenue for all electronic products enabled by chips and other components—such as computers, cell phones, network equipment, consumer electronics, medical instruments, automotive parts, and military systems—totals at least ten times as much, or $3 trillion dollars or more every year. The total supported employment in these fields reaches hundreds of millions of people, not to mention the countless more whose professions depend on the capabilities provided by such electronic products.

THE INFORMATION REVOLUTION

The proliferation of personal computers, smartphones, and the Internet has made it so that we no longer have to leave home to know everything that is going on in the world. Global news, sports, stock prices, and even the ability to instantly video chat face-to-face with long-distant loved ones or groups of close friends are available at our fingertips. If we want to collect information on any topic, search engines can provide us with answers from databases all over the world within a matter of seconds. Entertainment has become more and more of a screen-focused activity; tens of thousands of movies, shows, and segments of archival footage can be streamed and watched at any time, all at a very affordable cost. Hundreds of thousands of people can now simultaneously play video games together and compete in real time. All these massive amounts of Internet traffic are routed through "server farms" located all over the globe, and connected with optical fibers. Today, each server farm can consist of over one million microprocessors (incorporated into large numbers of individual servers) and over one million lasers to transmit data. It is no wonder that over 2.3 billion people worldwide now claim to have access to the Internet.

On the mobile side of the technology, over 4.6 billion mobile phones are currently in use—that's equivalent to 75 percent of the world's entire population. Strikingly, one can reach any of these people from any place on earth with signal coverage, and at any time by just gently tapping on a screen with the right phone number on it. The wireless phone signals are transmitted and received through the air via numerous base stations, which are themselves connected by fiber-optic cables feeding into the massive global switching network or orbiting satellites. Think of what an advance this is from the slow-moving time of the telegraph, when messages had to be relayed over wire and through multiple human conduits, or the still slower-paced time before that! Many of the dreams of science fiction and fantasy have actually been surpassed by modern technology.

It is generally agreed that, since the close of the Middle Ages, there have been three technologically inspired global revolutions that have shaped society and human behavior. The first was the original Industrial Revolution born in England in the eighteenth century. James Watt's steam engine provided the prime power that removed the fundamental physical limitation of the strength of humans, horses, or oxen. This allowed people to build machines that could move faster and more powerfully than ever before, and it also led to the mechanization of many important functions in manufacturing, transportation, and mining. At its core, this revolution was physical. It significantly improved human productivity and at the same time created innumerable social changes. No longer were serfs necessarily subservient to landowners and landowners to religious elites. The Industrial Revolution abruptly altered the lifestyle and social structure that mankind had lived under for thousands of years.

In the last half of the nineteenth century, the world went through the Second Industrial Revolution as a result of major breakthroughs in communications, energy, and transportation technologies. In communications, the telegraph, telephone, and wireless totally transformed the way people interfaced, annihilating the traditional barrier of physical distance between people. In energy, the emergence of the electrical power grid provided a wide-reaching, efficient, and clean network far superior to old wood- or coal-burning steam engines. Electrification and electrical lighting had particularly lasting impacts on our lives. In transportation, affordable automobiles powered by internal combustion engines and urban mass-transit networks revolutionized the range of personal mobility—no longer were people largely pigeonholed within one geographic area. The United States was at the center stage of this Second Industrial Revolution, and new electromagnetic technologies played an important role in its emergence.

Beginning in the mid-1970s, humanity has entered a Third Industrial Revolution that is information-oriented and enabled by electronics technology and software-oriented information technology. Even to this day, we are still within it. As this transformation has highlighted the extraordinary value of information and knowledge, the Third Industrial Revolution is also commonly known as the Information Revolution. The enabling technology for the First Industrial Revolution was the steam engine and the core technologies driving the Second Industrial Revolution were advancements like electrical generators and motors, light bulbs, telegraphy, the telephone, and the internal combustion engine. The cornerstones of the Information Revolution are computers (mainframe and personal), the Internet, smartphones, mass memory devices, and the broadband fiber-optic communications backbone, as well as various information technologies such as digital signal processing algorithms, database management, and network and systems software. In the Information Revolution, electronics technology plays the lead role.

During the forty-year span from 1970 to 2010, the information processing capability of microprocessor chips, data storage capacity of hard disks and memory chips, and data transmission rates in communication systems have also experienced million-fold improvement. Never before in history has there been such an enormous convergence of so many important and complementary technologies. Together they have pushed the Information Revolution to new highs and the momentum is still going strong.

GLOBALIZATION

As electronics-focused industries continue to evolve, the industry structures and supply chains also change. Silicon chips, LEDs, LCDs, and many other technologies were all invented in America, but in the last few decades, the manufacturing base of these products has shifted to Asia. To build and maintain state-of-the-art semiconductor and LCD manufacturing facilities requires extremely large capital investments; it also presents complex management and operational challenges. Based on contemporary business philosophy in the

West, where short-term return on investment is the key metric, American companies will likely continue to look for outsourcing opportunities to avoid capital investment, provided multiple qualified seller-suppliers exist. For now, they do exist indeed, and principally in Asia.

Beginning in the late 1960s, the Japanese government began to look beyond traditional businesses for opportunities to build new industries and stimulate economic growth. The semiconductor industry was one of the key industries that the Japanese targeted. With encouragement and assistance from the government, the Japanese semiconductor industry grew spectacularly in the 1970s and 1980s. Japanese companies showed a particularly high level of manufacturing competence and a willingness to work for tiny margins. In the face of this kind of competition, even an industry leader like Intel was forced to back away and find a new line of products.

Trying to copy Japan's success, South Korea, Taiwan, and Singapore all launched government-assisted efforts in the mid-1970s to build their own electronics industries. This strategy was followed by China in the 1990s. Their initial emphasis was on chip foundries, hard disk memory production, advanced electronic packaging, assembly and testing; later, their plans were extended to include LCD manufacturing as well. Contrary to American and European companies, whose management culture tends to focus on risk aversion, Asian companies tend to be more than willing to make large capital outlays to invest in manufacturing, including paying hefty royalties to license foundational know-how. To these companies—and the governments that provide them with key economic and political buttressing—the low profit margins in manufacturing pale in importance to creating large numbers of manufacturing jobs for local populations.

The majority of American and European companies are only too happy to see companies in other countries spend money to build manufacturing facilities. By purchasing high-quality, subcontracted manufacturing services, Western companies can focus their resources on more highly profitable segments of the product value chain, such as research and patent generation, product definition and design, branding and marketing, and channel management. This strategy, which has come to dominate the Western business landscape over the last thirty years, has enabled these companies to operate with high profit margins and achieve very high returns on investment. Despite this, however, it remains to be seen whether or not this strategy is sustainable over the long term. The answer is unclear, especially in the face of job losses in the well-paid manufacturing sector, which has never been a key metric for performance in the business and financial worlds, but may end up having long-term social and political repercussions.

Another strategic reason for Asian companies' and governments' willingness to make sizeable technological investments was the notion that they might first build a manufacturing foundation with licensed intellectual properties, and then, through continuous upgrading of capabilities and work force training, move up the value chain to challenge previously established companies with their own products. As core manufacturing capabilities migrated en masse to Asia, many Asian companies have begun to gain momentum to close the technical gap, particularly in hardware technology. For instance, Asian companies have an overwhelmingly dominant position over the West in LCDs. It is unlikely that the West will ever regain its historic lead in the field when the next generation of display technologies enters the market.

As the world becomes ever flatter, the winners will be those who manage to develop new businesses, out-innovate their competitors, and stay nimble. Today, the tremendous success of Apple Computer with their iPhone and iPad, to say nothing of their wildly lucrative App store, is a good illustration of the value of innovation. Of course, time doesn't stand still; the market will evolve. The smartphone is already rapidly becoming a commodity, and Asian companies, such as Samsung, are significantly eroding Apple's market position. In mobile communication and computing products, many chip design companies from traditionally different markets, such as those producing electronic games and cell phones, are now successfully challenging Intel's position on the ultra-low power microprocessor chips for such platforms. Who will be the next Apple or the next Intel? Which companies will out-innovate their competitors, not just in electronics or software, but with new technologies and in new industries that are yet to be discovered and developed? In reality, these companies are really synonymous with the outstanding individuals who founded them and provided the vision and driving force behind their success. Founders of companies such as Microsoft, Apple, Google, Amazon, Facebook, Tesla Motors, and many others are continuing the tradition set down by the *conquerors of the electrons* featured in this book. Who will be on the next list? They can be from all over the world. For leaders of the world engaged in this global competition, their challenge is to create and sustain an environment that can encourage and nourish such people and their creativity, both in its infrastructure and, more importantly, in its cultural setting.

LOOKING AHEAD

For almost fifty years, chip technology has consistently and exponentially progressed in accordance with Moore's Law. However, it appears that the physical limit of transistor scaling is about to be reached. This statement is not meant to be an over-dramatization; it is simply scientific fact. The driving force of the success of Moore's Law has been the ability to proportionately shrink the size of transistors, allowing chip performance to improve while reducing chip cost. Thanks to incredible, continuous technological advances, the minimum physical dimension of a transistor has shrunk in size from 10,000 nanometers in 1970 to a mere twenty-six nanometers in 2012. This is a nearly 400-fold shrinkage in linear dimension. Researchers are hard at work to shrink the transistor even further by using the third dimension. By 2020, these 3-D transistors with vertical fin-like structures, known as FinFETs, will likely shrink the minimum dimension to about seven nanometers, which corresponds to the spacing of just ten silicon atoms! It would appear that, were transistors to shrink any further from that point, there would be no room left for physical material at all!

What is waiting in the wings to replace silicon chips? There are many promising technologies in development, but no one can be sure which will take center stage. Some believe that nanomaterials such as graphene—a single layer of carbon atoms arranged in a perfectly arrayed honeycomb lattice—will play a hitherto inconceivable role in enabling new generations of devices. Others believe in molecular electronics, quantum computing, spintronics, or other esoteric ideas. The real breakthrough may not even lie in finding an ever-smaller transistor replacement: other factors, such as system architecture, may have a bigger impact in future computer systems. Picture the human brain with its complex network of neurons and synapses. The organic brain can outperform a supercomputer in solving many classes of problems, and it consumes much less energy. Yet, the electrical conduction speed of ions in a neural network is much slower than that of electrons flowing through a chip. What makes the brain so efficient? And what can engineers learn from that? Research on technologies such as neural networks has only scratched the surface of the subject before running into a new stone wall. Does the future hold an illuminating breakthrough in electro-bioengineering? It will come, but it is difficult to predict the proper path to the right exit of this enormously complex labyrinth.

In time, all technologies, like all people, reach a mature stage accompanied by the slowing of new progress. The field of electronics has experienced fifty years of continuous development at dazzling speed and has reached an extremely high state of complexity. Chips as we currently know them will probably cease making the accelerated gains prophesized by Moore's Law; however, neither design architecture nor software engineering suffer under these same kinds of physical limits, and we can thus still expect that ever more complex and powerful chip products will endlessly appear, providing for the continuing benefit and advancement of the human race.

Though silicon chip fabrication technology may slow down, the speed of optical fiber communications and the density of memory storage in hard disks will continue to grow. The bandwidth, too, of wireless communications will continue to expand as advances in CMOS technology push wireless chips to operate at ever higher frequencies and with ever more powerful signal processing capabilities. MEMS technology will find new uses, such as in miniaturized sensors and actuators in the biomedical industry. Imagine: tiny integrated MEMS sensors could be implanted onto arterial walls to monitor blood sugar and cholesterol levels in humans in real time. Signals from these MEMS chips inside the body could be communicated to a smartphone the owner is carrying. The information could be processed by a computer embedded in the smartphone or through a Cloud-based personal diagnostic center. If action is needed, commands could be sent to the actuator incorporated on the implanted MEMS chip to release the right amount of insulin—or any other drugs pre-stored on the chip—the moment they are needed. And what if no battery replacement were ever needed? Electricity to power the MEMS chip could be harvested from the wearer's own heartbeats! While applications such as these are mere dreams at the present, given another ten or twenty years, widespread and safe use of these devices may seem less and less like a far-fetched fantasy.

The fates of other technologies, too, remain to be determined. LCDs still enjoy healthy development, but new LED displays made from organic semiconductors, OLEDs, could take over commercial markets. OLEDs can themselves emit light, giving them a fundamental advantage in power consumption over conventional LCDs with backlighting. These devices may someday be manufactured with ink-jet printing on flexible surfaces, and integrated matrix switch arrays might also be fabricated on the same organic semiconductor films. If production costs for reliable OLEDs can be brought below those for the LCD—and there's no reason to believe they won't in the future—the so-revolutionary LCD displays developed by Sharp could become extinct, dinosaurs just as we think of CRTs today. OLED technology, in turn, may have a significant impact on other fields, like low cost, large scale solar energy conversion. There are other disruptive display technologies on the horizon, such as 3-D volumetric displays based on lasers; however many fundamental technical and economical hurdles still need to be overcome before they can be serious candidates.

And while we're on the topic of power generation based on OLED solar cells, we ought not to forget about batteries, the very technology that started off this whole electronic revolution. Volta introduced his "pile" to the world at the dawn of the 1800s, some two hundred-plus years ago. Now, lithium-ion batteries—compact, potent, and found in high-end consumer electronics from cameras to cell phones, and now, even high performance cars—are king. Surely, the lithium-ion battery will continue its steady evolutionary progress, but is that the ultimate solution? The basic structure of the battery has not changed since Volta, which, if you stop to ponder it, is amazing. Conventional battery chemistry is largely limited by the few potential combinations of electrode and electrolyte materials. But maybe someday, new nanomaterials, such as graphene-related two-dimensional materials, will be cleverly employed to enable a generation of high-density energy storage devices barely conceivable under our modern levels of understanding. Though hard to imagine, such a future is certainly possible. After all, mankind's journey of harnessing the power of electrons has been marked, more than anything else, by unexpected surprises and accidental discoveries. It is certainly possible that one of these unforeseeable breakthroughs may be just around the corner.

Striding forward in the adolescence of the twenty-first century, we find ourselves living at the height of the Information Age. But what will come next? The Green Energy Breakthrough? The Health Care Revolution? Mankind's needs for energy are just as boundless and inexhaustible as his demands for healthy living and environmental protection, and continuous progress in electronics technology and information sciences will assuredly play a major role in shaping our future, whatever that future may be.

APPENDIX I: FURTHER READING

GENERAL HISTORY OF ELECTRONICS

Brinkman, W. F., and D. V. Lang. "Physics and the Communication Industry." *Review of Modern Physics* 71, S480, March, 1999.

Brodsky, Ira. *The History of Wireless: How Creative Minds Produced Technology for the Masses.* Telescope Books, 2008.

Buchwald, Jed, and Andres Warwick, eds. *History of the Electron: The Birth of Microphysics.* MIT Press, 2001.

Bunch, Bryan, with Alexander Hellemans. *The History of Science and Technologies.* Houghton Mifflin, 2004.

Campbell-Kelly, Martin, and William Aspray. *Computer: A History of the Information Machine.* Perseus Books Group, 1996.

Chandler, Alfred, Jr. *Inventing the Electronic Century: The Epic Story of Consumer Electronics and the Computer Industry.* Simon & Schuster, 2001.

Gribbin, John. *The Scientists.* Random House, 2002.

Pearson, Gerald, ed. "Historical Notes on Important Tubes and Semiconductor Devices." *IEEE Bi-Centennial Issue on Electronic Devices.* IEEE, July 1976.

Rowland, Wade. *Spirit of the Web: The Age of Information from Telegraph to Internet.* Somerville House, 1997.

Computer History Museum archive. http://www.computerhistory.org

Electronic history from IEEE website. http://www.ieeeghn.org

Historical events, with emphasis on battery technology. http://www.mpoweruk.com/history.htm#leyden

Nobel Prize lectures. www.nobelprize.org/nobelprizes/physics

Silicon Genesis Oral history. http://silicongenesis.stanford.edu/complete_listing.html

ELECTROMAGNETISM

Scientific Basis

Hirschfeld, Alan. *The Electric Life of Michael Faraday*. Walker, 2006.
Ludwig, Charles. *Michael Faraday, Father of Electronics*. Herald, 1978.
Mahon, Basil. *The Man Who Changed Everything: The Life of James Clerk Maxwell*. John Wiley and Sons, 2004.

The Telegraph

Abbate, Janet. *Inventing the Internet*. MIT Press, 1999.
Silverman, Kenneth. *Lightning Man: The Accursed Life of Samuel F. B. Morse*. Alfred Knopf, 2003.
Standage, Tom. *The Victorian Internet*. Penguin Putnam, 1998.

The Telephone

Grosvenor, Edwin, and Morgan Wesson. *Alexander Graham Bell: The Life and Times of the Man Who Invented the Telephone*. Harry Abrams, 1997.
Shulman, Seth. *The Telephone Gambit: Chasing Alexander Graham Bell's Secret*. Norton, 2008.
AT&T's "Unnatural Monopoly." http://www.cato.org/pubs/journal/cjv14n2-6.html
Fagen, F. D. Telephone History: "A History of Engineering and Sciences at Bell Systems." Bell Labs, 1975.
History of Bell Systems. http://www.porticus.org/bell/bellsystem_history.html#Year%20of%20Decision
PBS. "American Experience: The Telephone—Elisha Gray." http://www.pbs.org/wgbh/amex/telephone/peopleevents/pande02.html
Vail, Ted. http://www.cato.org/pubs/journal/cjv14n2-6.html

Wireless Telegraphy

Weightman, Gavin. *Signor Marconi's Magic Box*. Perseus Books, 2003.

Lighting and Electrification

Cheney, Margaret. *Tesla*. Prentice-Hall, 1981.
Israel, Paul. *Edison: A Life of Invention*. John Wiley & Sons, 1998.
Jonnes, Jill. *Empires of Light: Edison, Tesla, Westinghouse, and the Race to Electrify the World*. Random House, 2003.

Millard, Andre. *Edison and the Business of Innovation.* Johns Hopkins Press, 1990.
Edison. http://www.pbs.org/wgbh/amex/edison/timeline/index_2.html
John Jenkins' Spark Museum. http://www.sparkmuseum.com
Siemens history. http://www.siemens.com/history/de/index.htm

VACUUM ELECTRONICS

Electrons and the X-ray

Kevles, Bettyann Holtzmann. *Naked to the Bone: Medical Imaging in the 20th Century.* Rutgers University, 1997.
Thompson, J. J. http://nobelprize.org/nobel_prizes/physics/laureates/1906/thomson-bio.html
Rontgen and the X-ray. http://nobelprize.org/nobel_prizes/physics/laureates/1901/rontgen-bio.html

Vacuum Triode

Zouary, Maurice. *De Forest: Father of the Electronic Revolution.* (ISBN: 1-58721-449-0.) William H. Allen, 1995.
De Forest, Lee. http://www.pbs.org/wgbh/aso/databank/entries/btfore.html

Radio

Brodsky, Ira. *The History of Wireless: How Creative Minds Produced Technology for the Masses.* Telescope Books, 2008.
Armstrong, Edwin. http://world.std.com/~jlr/doom/armstrng.htm
Early radio. http://earlyradiohistory.us/index.html
Spark Museum. http://www.sparkmuseum.com/RADIOS.HTM
Sarnoff, David. http://www.museum.tv/eotvsection.php?entrycode=sarnoffdavi

Television

Fisher, David, and Marshall Jon Fisher. *Tube: The Invention of Television.* Harvest, 1996.

Radar

Boot, H. A. H., and J. T. Randall. "Historical Notes on the Cavity Magnetron." *IEEE Transactions on Electronic Devices*, Vol. 23, No. 7, 1976.
Buderi, Robert. *The Invention That Changed the World.* Touchstone, 1996.

The Electronic Computer

Campbell-Kelly, Martin, and William Aspray. *Computer: A History of the Information Machine.* Perseus Books Group, 1996.

Computer History. http://www.computersciencelab.com/ComputerHistory/History.htm

University of Pennsylvania history on ENIAC. http://www.upenn.edu/almanac/v42/n18/eniac.htm

SOLID STATE ELECTRONICS

Bell Labs and the Semiconductor

Gertner, Jon. *The Idea Factory—Bell Labs and the Great Age of American Innovation.* The Penguin Press, 2012.

AT&T History. http://www.corp.att.com/history/milestones.html

Invention of the Transistor

Gilder, George. *Microcosm: The Quantum Revolution in Economics and Technology.* Touchstone, 1987.

Riordan, Michael, and Lillian Hoddeson. *Crystal Fire: The Invention of the Transistor and the Birth of the Information Age.* Norton, 1997.

Shockley, W. "The Path to the Conception of Junction Transistor." *IEEE Transactions on Electronic Devices*, Vol. 23, No. 7, 1976.

Shurkin, Joel. *Broken Genius: The Rise and Fall of William Shockley, Creator of the Electronic Age.* Macmillan, 2006.

Teal, G. K. "Single Crystals of Germanium and Silicon: Basic to the Transistor and Integrated Circuit." *IEEE Transactions on Electronic Devices*, Vol. 23, No. 7, 1976.

Nobel lectures by Shockley, Bardeen, and Brattain. www.nobelprize.org/nobelprizes/physics

PBS. Transistor. http://www.pbs.org/transistor/

Commercialization

Bell Licensing the Transistor. http://www.pbs.org/transistor/background1/events/symposia.html

Silicon Valley and the Chips

Kaplan, David A. *The Silicon Boys.* Perennial, 2000.

Moore, G. M. "The Role of Fairchild in Silicon Technology." *Proceedings of the IEEE*, Vol. 86, No. 1 (January 1998): 53–62.

Reid, T. R. *The Chip*. Random House 1985, 2001.

Silicon Genesis. (Oral interview of many key figures during the period, including Gordon Moore, Jay Last, Morris Chang, Ted Hoff, Federico Faggin, Charlie Spork, Jerry Sanders, Alfred Yu, Les Hogan, and Arthur Rock.) http://silicongenesis.stanford.edu/complete_listing.html

Silicon Valley History. http://nobelprize.org/nobel_prizes/physics/articles/lecuyer/index.html

Chip Technology

Berlin, Leslie. *The Man Behind the Microchip: Robert Noyce and the Invention of Silicon Valley*. Oxford, 2005.

Boyle, W. S., and G. E. Smith. "The Inception of the Charge Coupled Device." *IEEE Transactions on Electronic Devices*, Vol. 23, No. 7, 1976.

Goldstein, Andrew, and William Aspray. "Social Construction of the Microprocessor: A Japanese and American Story." In *Facets: New Perspectives on the History of Semiconductors*. New Brunswick: IEEE Center for the History of Electrical Engineering, 1997, pp. 215–267.

Riordan, Michael. "The Silicon Dioxide Solution: How Physicist Jean Hoerni Built the Bridge from the Transistor to the Integrated Circuit." *IEEE Spectrum* (December 2007): pp. 44–50.

Yu, Albert. *Create the Digital Future: The Secrets of Consistent Innovation at Intel*. The Free Press, 1998.

First Microprocessor. (Oral interview by Faggin and Shima.) http://www.ieeeghn.org/wiki/index.php/Oral-History:Federico_Faggin; http://www.ieeeghn.org/wiki/index.php/Oral-History:Masatoshi_Shima

Gordon Moore: Accidental Entrepreneur. http://nobelprize.org/nobel_prizes/physics/articles/moore/index.html

矽說台灣-台灣半導体產業傳奇, 張如心, 天下文化 (The Legend of the Silicon Industry in Taiwan, "Silicon Talking in Taiwan: Taiwan Semiconductor Industry Legend," Chang [Chang Rhu-Shing], Commonwealth Publication, Taiwan, June, 2006.)

Chang, Morris. 張忠謀自傳上冊, 天下文化, 2001 (Autobiography of Morris Chang [Chang Tsung-Mou], Part I, Commonwealth Publication, Taiwan, 2001)

Electro-Optical Technologies

LED

Welker, H. J. "Discovery and Development of III-V Compounds." *IEEE Transactions on Electronic Devices*, Vol. 23, No. 7, 1976.

Zheludev, N. "The Life and Times of the LED: A 100-Year History." *Nature Photonics* 1 (4) (2007): 189–192. doi:10.1038/nphoton.2007.34

Semiconductor Laser

Agrawal, G. P., and N. A. Dutta. *Semiconductor Lasers*. Van Nostrand Reinhold, New York, 1993.

Schawlow, A. L. "Masers and Lasers." *IEEE Transactions on Electronic Devices*, Vol. 23, No. 7, 1976.

The Laser & Townes. http://www.bell-labs.com/history/laser/

Fiber Optics

Hecht, Jeff. *City of Light: The Story of Fiber Optics*. Oxford University Press, 1999.

Kaminow, I. P., and T. L. Koch. *Optical Fiber Telecommunications*. Academic Press, San Diego, 1997.

Short history of fiber optics. http://www.sff.net.people/Jeff.Hecht/history.html

Liquid Crystal Displays

Heilmeier, G. H. "Liquid Crystal Displays: An Experiment in Interdisciplinary Research That Worked." *IEEE Transactions on Electronic Devices*, Vol. 23, No. 7, 1976.

Kawamoto, Hirohisa. "The History of Liquid Crystal Displays." *Proceedings of the IEEE*, Vol. 90, No. 4, April 2002.

IEEE Review Article on LCD. http://www.ieee.org/portal/cms_docs_iportals/iportals/aboutus/history_center/LCD-History.pdf

APPENDIX II: SUMMARY OF KEY "CONQUERORS OF THE ELECTRON"

Name	*Life Span*	*Age*	*Nationality*	*Main Contribution*
Electromagnetism				
William Gilbert	1544–1603	59	Great Britain	First to use Scientific Method to study electromagnetism; coined the word "electron"
Otto von Guericke	1602–1686	84	Germany	Invented Guericke Sphere to generate static electrical charge; also invented vacuum
Stephen Gray	1666–1736	70	Great Britain	Discovered conductors and insulators for electricity
Charles du Fay	1698–1739	41	France	Discovered positive and negative charges
Pieter van Musschenbroek	1692–1761	69	Netherlands	Invented Leyden Jar to store electric charge
Jean Nollet	1700–1770	70	France	Conducted first electric conduction experiment through human bodies
Benjamin Franklin	1706–1790	84	USA	Proved lightning was due to electric discharge in clouds; invented lightning rod

(continued)

Name	*Life Span*	*Age*	*Nationality*	*Main Contribution*
Charles de Coulomb	1736–1806	70	France	Experimentally developed Coulomb's law for electrostatic interaction
Luigi Galvani	1737–1798	61	Italy	Key figure in the study of "Animal Electricity"
Alessandro Volta	1745–1827	82	Italy	Invented battery (Voltaic Pile)
André Ampère	1775–1836	61	France	First to use mathematics to describe relationship between current and magnetism
Hans Oersted	1777–1851	74	Denmark	First observed the linkage between electrical current and magnetism
Humphry Davy	1778–1829	51	Great Britain	Electrochemist; discovered many alkaline metals; demonstrated electrical arcing
William Sturgeon	1783–1850	67	Great Britain	Discovered practical electromagnets
Michael Faraday	1791–1867	76	Great Britain	Great experimentalist; demonstrated concept of motor, generator, induction & others
Joseph Henry	1797–1878	81	USA	Early American scientist in electromagnetism; demonstrated relay concept
James Clerk Maxwell	1831–1879	48	Great Britain	Greatest theorist; Maxwell equations provided the foundation of all electromagnetism
Heinrich Hertz	1857–1894	37	Germany	First to generate and receive electromagnetic waves
Telegraph				
Samuel F. B. Morse	1791–1872	81	USA	Inventor of telegraph
William Cooke	1806–1879	73	Great Britain	Inventor of the Cooke/ Wheatstone telegraph in Great Britain

Name	Life Span	Age	Nationality	Main Contribution
Alfred Vail	1807–1859	52	USA	Key member of Morse's team; may be the inventor of the Morse Code
Werner von Siemens	1816–1892	76	Germany	Inventor of the Point Telegraph; founder of the Siemens company
Lord Kelvin (William Thomson)	1824–1907	83	Great Britain	Great physicist/ engineer; responsible for the successful Transatlantic Telegraph Cable
Telephone				
Gardiner Greene Hubbard	1822–1897	75	USA	The force behind Bell's "invention" of the telephone & the launching of the business
Elisha Gray	1835–1901	66	USA	Conceptual inventor of transmitting voice and music on wire
Theodore (Ted) Vail	1845–1920	75	USA	The business genius that built Bell Telephone and AT&T
Alexander Graham Bell	1847–1922	75	USA (Scotland)	Well known as inventor of the telephone (though real case is controversial)
Wireless Telegraphy				
John Ambrose Fleming	1849–1945	96	Great Britain	Developed vacuum diode for wireless receiver applications
Heinrich Hertz	1857–1893	36	Germany	First to discover, generate, and receive radio waves
Guglielmo Marconi	1874–1937	63	Italy	Pioneered wireless telegraphy and built a global company
Lighting and Electrification				
Werner von Siemens	1816–1892	76	Germany	Founder of the Siemens company; refined telegraph, dynamo, and loud speakers

(continued)

Name	*Life Span*	*Age*	*Nationality*	*Main Contribution*
Zenobe Gramme	1826–1901	75	Belgium	Inventor of practical motor and generator; demonstrated electrical power transmission
J. P. Morgan	1837–1913	76	USA	Key capitalist; revived GE and AT&T
George Westinghouse	1846–1914	68	USA	Founder of Westinghouse Electric; builder of the AC power system
Thomas Alva Edison	1847–1931	84	USA	Great inventor involving telegraph, microphone, phonograph, lighting, power grid, & others
Nikola Tesla	1856–1943	87	USA (Serbia)	Technical genius; inventor of AC motor and architect of multiphase AC power system
Vacuum Electronics				
Heinrich Geissler	1814–1879	65	Germany	Developed high vacuum technology; observed gas discharge phenomenon
William Crookes	1832–1919	87	Great Britain	Discovered cathode ray
Wilhelm Roentgen	1845–1923	78	Germany	Discovered X-ray
John Ambrose Fleming	1849–1945	96	Great Britain	Developed vacuum diode for wireless detection
Karl Ferdinand Braun	1850–1918	68	Germany	Developed CRT; discovered metal-PbS rectification phenomenon; contributor in wireless telegraphy
J. J. Thomson	1856–1940	84	Great Britain	Identified electron
Lee De Forest	1873–1961	88	USA	Invented vacuum triode
Radio				
Reginald Fessenden	1866–1932	66	Canada	First to conceive and demonstrate radio broadcast concept

Name	Life Span	Age	Nationality	Main Contribution
Edwin Armstrong	1890–1954	64	USA	Contributor in core radio technology, including oscillator, Superheterodyne detection, and FM
David Sarnoff	1891–1971	80	USA (Russia)	Key visionary and leader in developing radio broadcasting as a new mass media
Television				
Paul Nipkow	1860–1940	80	Germany	Created first mechanical scanner for imaging applications
John Baird	1888–1946	58	Great Britain	First to demonstrate a television with Nipkow scanner
Vladimir Zworykin	1888–1982	94	USA (Russia)	Developed the successful television technology at RCA
David Sarnoff	1891–1971	80	USA (Russia)	Masterminded the first television broadcast
Philo Farnsworth	1906–1971	65	USA	First to demonstrate an all-electronic television using CRT
Radar				
Robert Watson-Watt	1892–1973	81	Great Britain	Leader of the Chain Home coastal defense radar system
John Randall	1905–1984	79	Great Britain	Inventor of the resonant cavity magnetron that enabled airborne radar
Henry Boot	1917–1983	66	Great Britain	Co-inventor of the resonant cavity magnetron
Robert J. Dippy	1912–?		Great Britain	Technical leader for the GEE radio navigation system that ultimately led to the GPS system
Computer				
Charles Babbage	1791–1871	80	Great Britain	Inventor of the "Difference Engine," origin of complex mechanical computers

(continued)

Name	*Life Span*	*Age*	*Nationality*	*Main Contribution*
George Boole	1815–1864	49	Great Britain	Mathematician that developed binary system for computing and logic
John von Neumann	1903–1957	54	USA (Hungary)	Key contributor to modern computer architecture
John Atanasoff	1903–1995	92	USA	Pioneer of the binary electronic computer concept
John Mauchly	1907–1980	73	USA	Primary force behind the ENIAC all-electronic computer, the origin of modern computer
J. Presper Eckert	1919–1995	76	USA	Co-leader of the ENIAC project
Transistor				
Walter Schottky	1886–1976	90	Germany	First applied solid-state physics principles to interpret metal-semiconductor rectification
Mervin Kelly	1894–1971	77	USA	Research director of Bell Labs, the cradle of semiconductor technology
Russell Ohl	1898–1987	89	USA	Discovered the p-n junction
Arnold Beckman	1900–2004	104	USA	Founder of Beckman Instruments; investor in first semiconductor company in Silicon Valley
Walter Brattain	1902–1987	85	USA	Inventor of the first point-contact transistor
Alan H. Wilson	1906–1995	89	Great Britain	Theoretical physicist who first applied solid-state physics to the study of semiconductor
Gordon Teal	1907–2003	96	USA	Developed single-crystal semiconductor material; made first junction transistor, and the Si transistor

Name	Life Span	Age	Nationality	Main Contribution
John Bardeen	1908–1991	83	USA	Inventor of the first point-contact transistor
Masaru Ibuka	1908–1997	89	Japan	Co-founder of Sony
William Shockley	1910–1989	79	USA	Inventor of the junction transistor & its theory; key figure in launching Silicon Valley
Jack Morton	1913–1971	58	USA	Leader of early transistor manufacturing team at Bell Labs; later its research director
Pat Haggerty	1914–1980	66	USA	President of Texas Instruments; led the company into semiconductor business
Akio Morita	1921–1999	78	Japan	Co-founder of Sony
Silicon Chips				
Gene Kleiner	1923–2003	80	USA	One of the founders of Fairchild Semiconductor; later founded Kleiner-Perkins (venture capital)
Jack Kilby	1923–2005	82	USA	Inventor of the first integrated circuit
Jean Hoerni	1924–1997	73	USA (Switzerland)	Developed the planar process that enabled the monolithic integrated circuit, or chips
Willard Boyle	1924–2011	87	Canada	Co-inventor of CCD
Arthur Rock	1926–		USA	Founding figure of venture capital in Silicon Valley; early investor in Fairchild, Intel, and Apple
Robert Noyce	1927–1990	63	USA	Founder of Fairchild Semiconductor & Intel; inventor of silicon chips; leader of US semiconductor industry

(continued)

Name	*Life Span*	*Age*	*Nationality*	*Main Contribution*
Gordon Moore	1929–		USA	Founder of Fairchild Semiconductor and Intel; originator of "Moore's Law"
George Smith	1930–		Canada	Co-inventor of CCD
Morris Chang	1931–		USA (China)	Founder of Taiwan Semiconductor Manufacturing Company (TSMC); started the pure-play foundry model
Carver Mead	1934–		USA	One of the key contributors in computer-aided chip design technology
Andy Grove	1936–		USA (Hungary)	Founder and key executive at Intel
Ted Hoff	1937–		USA	Developed the concept of the first microprocessor chip
Federico Faggin	1941–		USA (Italy)	Designed and produced the first microprocessor chip
LEDs, Fiber Optics, and LCD				
Heinrich Welker	1912–1981	69	Germany	First to synthesize artificial III-V semiconducting materials
Charles Townes	1915–		USA	Inventor of the laser and maser concept
Charles Kao	1933–		Great Britain (China)	Pioneer of fiber optic communication technology
George Heilmeier	1936–		USA	Pioneer of liquid crystal display technology
Shuji Nakamura	1954–		Japan	First to demonstrate high efficiency blue LEDs
Herb Kroemer	1928–		USA (Germany)	Developed heterojunction concept that enabled many new devices, including semiconductor lasers